MW01443508

Midwest Institute of Geosciences and Engineering

mige-web.org

© 2025

3rd Edition

Cover photo is the Eriksfjord Formation in Igaliku Greenland

Lab Reference Book
Classifying Igneous, Sedimentary, and Metamorphic Rocks

Steven D.J. Baumann, PG

PREFACE TO THIS CONSOLIDATED VOLUME

This book is the consolidation of the "Lab Reference Books" consolidated into one volume. It includes all three books in the series, "Lab Reference Books: Classifying Igneous Rocks, Lab Reference Books: Classifying Metamorphic Rocks, and Lab Reference Books: Classifying Sedimentary Rocks", in that order. But this edition is also more than that.

It must be noted that in this edition there are significant changes from just a consolidation. Herein there are many significant changes and additions, such as an entirely new fourth section. There are updated and new charts. This edition also has an index (p.314-324). From this point on, there will be no further updates of the three separate volumes. This reference book now functions as a single book.

Please remember that this book is an aid to identify rocks in a simple lab setting. **This book is NOT intended to be a teaching tool for basic geology courses. Nor is it meant to aid in origins and interpretations of rocks and their history.** Although some pages have been added that include interpretive classification schemes, just so you can compare them to other sources that still use interpretive systems.

Although this book can be used outside of the Midwest of North America (everything in the circle on p.222), its primary focus is the Midwest. Otherwise this book would have been way too thick.

AUTHOR CONTACT

You can email the author directly at:

proterozoicguy@gmail.com

This page is intentionally left blank

TABLE OF CONTENTS

ITEM	PAGE
Preface	3
Author contact	4
Table of contents	6
Lab Reference Book: Classifying Igneous Rocks	**13**
Introduction	15
Important notes when using this book	17
<u>Part 1: Overview and general characteristics of igneous rocks</u>	18
Wentworth Scale for Igneous rocks	19
Mineral hardness	21
Common igneous minerals	22
Unsaturated vs. saturated	24
Bowen's reaction series and tree	26
Simplified flow chart for classifying igneous rock	31
<u>Part 2: Characteristics of phaneritic igneous rocks</u>	32
IUGS QAPF plot for phaneritic (plutonic) rocks	33
Mafic and ultramafic phaneritic (plutonic) rock classification	35
Unique mafic and ultramafic rocks	39
Plagioclase vs. alkali-feldspar (K-spar)	42
Other key factors separating alkali-feldspar from plagioclase	45
Feldspathoids (Foids)	49
Diorite vs. gabbro	52
Volume calculations in phaneritic (plutonic) rocks	55
<u>Part 3: Characteristics of aphanitic igneous rocks</u>	58
Identifying aphanitic (volcanic) rocks	59
Porphyritic vs. amygdaloidal vs. pegmatitic textures	61
Important Notes on Aphanitic Rock Identification	64
IUGS QAPF plot for aphanitic (volcanic) rocks	65
Normalizing percentages in rocks (2nd order normalization)	67
Archean bs. Proterozoic igneous rocks (TTG, anorthosite, and granite)	70
Granophyre	71
Weathering patterns in identifying igneous rocks	72
Pyroclastic rocks	74
Pyroclastic rock classification diagram	76
Pyroclastic sediment classification diagram	77
Lamprophyre	78

ITEM	PAGE
Compositional, textural, rock name block diagram	79
Relative amounts of minerals in felsic v. intermediate v. mafic v. ultramafic rocks	80
Simplified classification comparison chart	82
Properties of common mafic minerals	84
Accepted pyroxene chemical subdivisions	89
Pyroxenite	92
Native elements and similar looking minerals	94
An example of classifying igneous rocks	98
Example of Naming an Igneous Rock	99
Worksheet for QAPF plot	101
Igneous Rock Samples Under High Magnification	102
Lab Reference Book: Classifying Metamorphic Rocks	**107**
Introduction	109
Part 1: Overview and general characteristics of metamorphic rocks	111
What is a metamorphic rock?	112
Metamorphic rock types	113
Grain shape of crystalline metamorphic rock	115
Crystalline size	116
Adapted Wentworth scale chart for crystalline size	117
Types and causes of metamorphism	118
Metamorphic facies	119
Metamorphic facies temperature and pressure diagram	120
Common metamorphic minerals	122
Hornfels and granofels problem	128
Mylonite problem	129
Characterizing bedding, foliations, lineations, and banding	131
Cleavage	132
Naming process	133
Part 2: Foliated metamorphic rocks	134
Part 3: Non-foliated metamorphic rocks	138
Sandite Classification Chart	140
Conglomerite	144
Ironite (Metamorphic Banded Iron Formation)	146
Stability fields of minerals in BIF	147

ITEM	PAGE
Ironite ternary plot	148
Marble ternary plot based off mineralogy	149
Part 4: Unique metamorphic rocks	150
Migmatite	151
Skarn	154
Metamorphic rocks under high magnification	156
Lab reference: Classifying sedimentary rocks	**161**
Introduction	163
Part 1: General sedimentary rock textures	164
Sedimentary rock fabric	165
Grain sorting	166
Intergrain relationships	167
Grain orientation / alignment	168
Grains	169
Grain angularity	170
Grain size	171
Wentworth scale	173
Particle size comparison	174
Grain shape	175
Grain distribution	176
Bedding	178
Cross beds	179
Bedding thickness terms	183
Color and mineralogy	184
Induration	185
Common sedimentary minerals	186
Sedimentary rock types based off traditional groupings	189
Part 2: Clastic rocks sedimentary rocks	190
What are clastic rocks?	191
Conglomerate	192
Sandstone	196
Mudstone	204

ITEM	PAGE
Part 3: Non clastic sedimentary rocks	211
Carbonates	212
HCl acid standards	213
Flowchart for Testing Carbonates	215
Crystalline Carbonates	216
Non Crystalline Grains	217
Carbonate ternary plot based off mineralogy	218
Lithological Carbonate Symbols	219
Chert	221
Evaporite	222
Banded iron formation (BIF)	223
Coal	227
Lithological Chertstone, BIF, Evaporite, and Coal Symbols	229
Sedimentary Rocks Under High Magnification	230
Lab reference book: Tables and charts applicable to all rock types or general information	**235**
Introduction	237
Earth's interior	238
Internal Mechanical Layers of the Earth and Their Percent Mass	239
Internal Mechanical Layers of the Earth and Their Percent Volume	240
A more realistic cross section of the interior of the Earth	241
Basic Mechanical Structures in Earth's Interior	242
Using density to calculate mass of Earth	243
Layers of Earth in percent volume	244
Acceptable color descriptions	245
Grain percentages based off volume	249
Basic properties of select rocks and minerals / mineraloids	252
Crystal systems	266
Density vs. specific gravity	268
Optics (for thin sections)	269
Relative particle sizes for comparison	271
International Mineralogical Association (IMA) abbreviations	273
Rock abbreviations	282

ITEM	PAGE
Molar thermodynamic data for common minerals at standard temperature and pressure, [298.15 K and 105 Pa (1 bar)]	284
Blank notes	286
Photos of select minerals	305
Unit Conversions	309
References	310
Index of figures	314
Index by subject	317

This page is intentionally left blank

This page is intentionally left blank

Lab Reference Book
Classifying Igneous Rocks

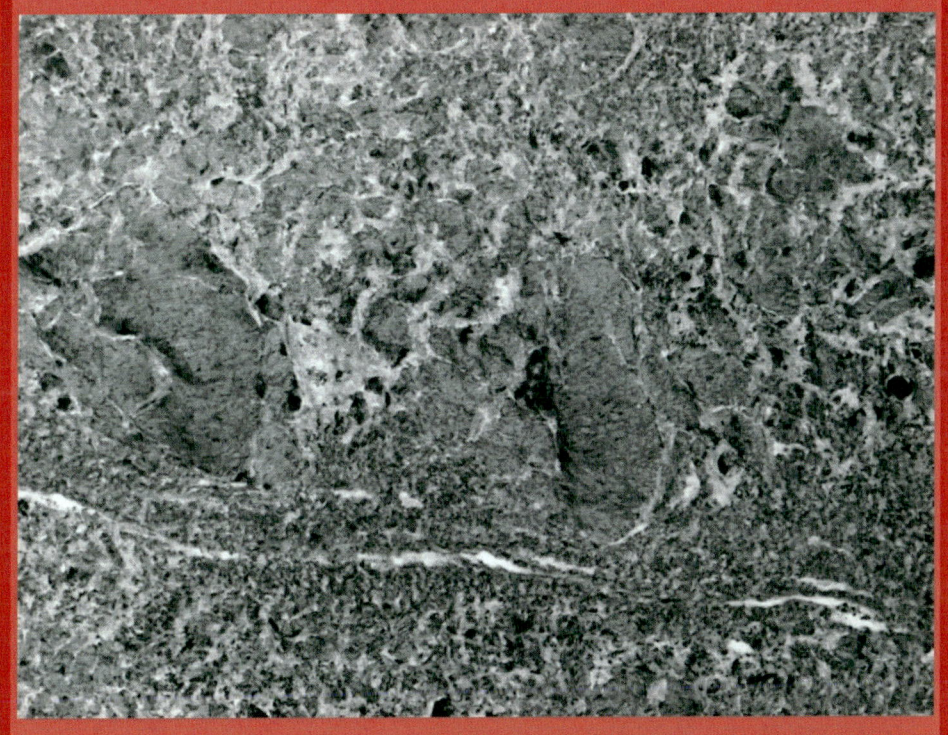

Steven D.J. Baumann, PG

Igneous cover photo (previous page) is of the interior walls of the "Memorial Library" on the campus of the University of Wisconsin in Madison. It was taken by the author as were all photos within this book.

INTRODUCTION

Rocks collected in the field are often brought back for more detailed analysis in a lab setting. The identification of rocks based off small scale texture and mineral composition is petrology. Identifying the composition of a rock is just as important as identifying its large scale structures in the field.

This book is a general guide for identifying igneous rocks visually in the lab. What do I mean by visually? I mean either through simple tests or through visual inspection that does *NOT* require thin sections. Most of us post-school, do not have easy access to academic institutions where we can cut thin sections and use polarized or non polarized light. You can still use a hand lens, magnifying glass, or a microscope.

Herein, I deal with major rock forming minerals in igneous rocks and some common secondary minerals. The primary focus of this book is on rocks common in the Midwest of North America. This book is not meant to be all inclusive, nor is it meant to include information for every possible rock collected from the field.

All photos were taken by the author. All drawings and diagrams are by the author. Adaptations are noted.

Problems with Identifying Rocks From a Photo

People will often send me photographs of rocks and ask, "what is it"? I don't do that. It is nearly impossible to identify a rock from a low resolution (or even high resolution) photo. Especially if the person has no clue where they picked it up from. Sometimes weathering patterns, or lack thereof, can be helpful. Usually, rock identification from photos should be avoided. Hands on experience is the only way to do it. Both in the field and in the lab.

Say, someone sends you the following photo, they want you to identify it, and all they can say about it is "I took the photo in Wisconsin". There's no scale, no orientation, nothing but the photo.

Wisconsin has igneous, sedimentary, and metamorphic rocks at the surface. We can't rule any of them out. The rock is light colored, apparently crystalline, and appears either bedded or cross cut with pink and dark gray bands.

This could be just about anything, except for a mafic volcanic or plutonic rock. If the light gray areas are carbonate, then this could be a dolostone (sedimentary) or even a marble (metamorphic). The light gray areas could be all plagioclase and quartz, making this an anorthosite to granodiorite cross cut with pink syenite and the dark band could be muscovite (igneous). It could even be a proto-quartzite (meta-sedimentary). To make things even worse, Wisconsin was glaciated within the last 2.5 million years. So if this is an erratic, it could have come from the Upper Peninsula of Michigan or even Ontario, complicating things even more!

I just wanted to point out to you, how difficult identifying rocks can be, and how experience is key. Hopefully this book will add to your body of knowledge and point out some things you may not have known.

Important Notes When Using This Book

This book is designed to aid in basic petrologic analysis in a laboratory setting. It is not intended to be used if detailed petrographic analysis is being conducted. If you have the ability to do thin sections, use an electron microscope, or run chemical analysis, this book can only be used as a generic guide. As a result, some things are kept out. For example this book does not contain a "total alkalis vs. silica" (TAS) diagram/plot, because TAS is for chemical analysis. When all you have is artificial light, hand lens, and a microscope, you can't determine exact chemical ratios.

Lamprophyres (p.31, 39, 40, 78) are another complicated subject as their classification system is still highly variable among users. So I will cover what they are and how to recognize them, but not how to classify them.

I also will not be dealing with carbonate igneous rocks as they are essentially non existent in the Midwest, except as minor components in rare intrusions.

All the detailed mineral data in this book was taken from the online version of the "Handbook of Mineralogy" (see references under: Anthony, et al.).

I hope you find this book informative and helpful! It's based not only off research but actual experience over the years as I attempt to classify rocks in a standard manner. There are still things to discover.

PART I:
OVERVIEW AND GENERAL CHARACTERISTICS OF IGNEOUS ROCKS

WENTWORTH SCALE FOR IGNEOUS ROCKS

The Wentworth scale is most often used for grain sizes in sedimentary rocks. However, it is a good logical standard to use during petrographic analysis. There is no engineering or agricultural equivalent.

On the next page the Wentworth scale is depicted with its phi (Φ) scale, millimeter (mm) scale, inch (") scale, and closest U.S. sieve size. Now you aren't going to be mechanically breaking down an igneous rock to run a sieve analysis. It is just included as a reference guide for estimating crystal sizes within a rock. To the right is the Wentworth scale for clastic rocks, e.g. boulders, cobbles, granules, sand, silt, and clay particles. It is also included for reference.

Megatite is a term herein defined as igneous crystals that are equal to or greater in size to 64 mm in the appropriate dimensional axis.

Modified Wentworth Size Scale Chart for Igneous Rocks

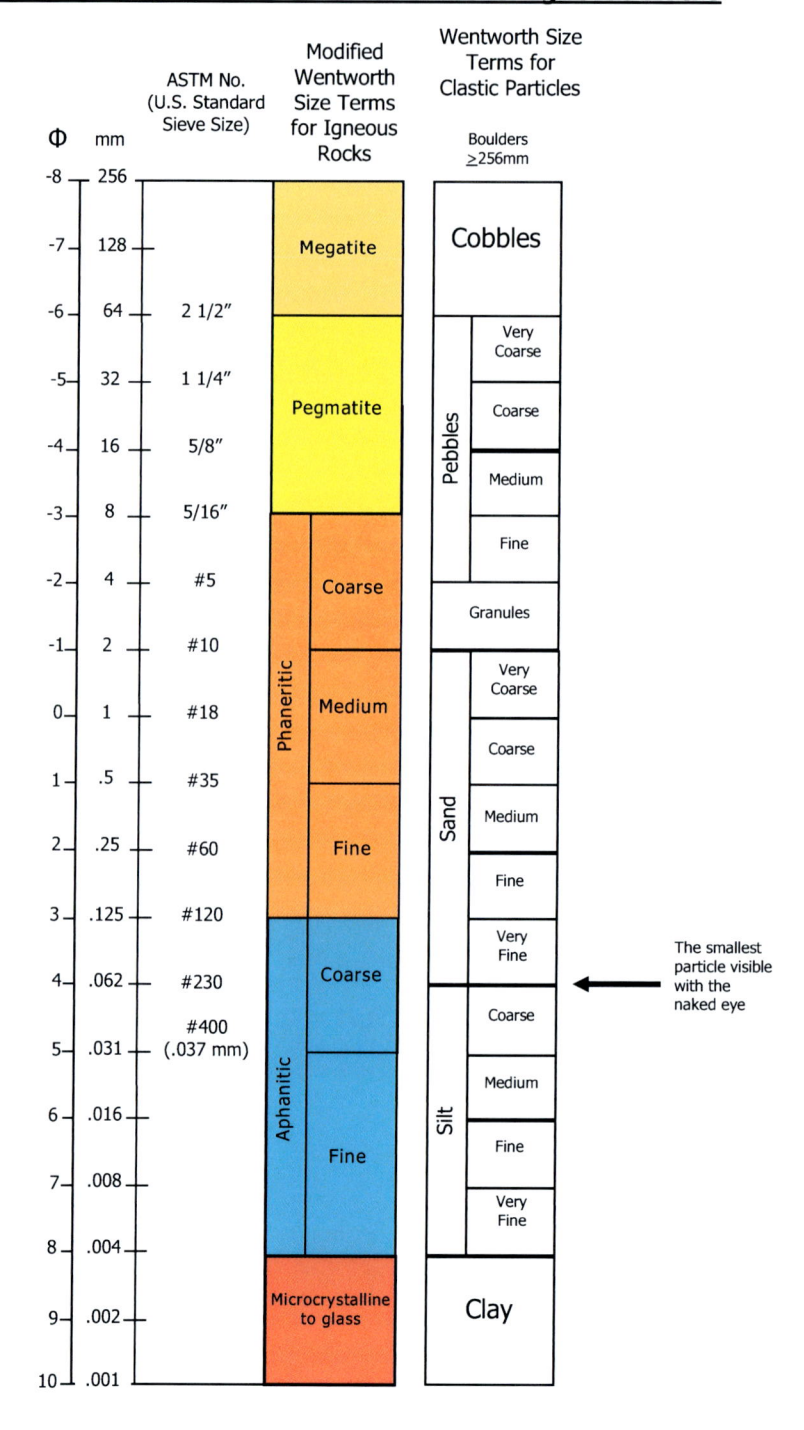

Φ	mm	ASTM No. (U.S. Standard Sieve Size)	Modified Wentworth Size Terms for Igneous Rocks	Wentworth Size Terms for Clastic Particles	
-8	256			Boulders ≥256mm	
-7	128		Megatite	Cobbles	
-6	64	2 1/2"	Pegmatite	Pebbles	Very Coarse
-5	32	1 1/4"			Coarse
-4	16	5/8"			Medium
-3	8	5/16"			Fine
-2	4	#5	Phaneritic Coarse		Granules
-1	2	#10		Sand	Very Coarse
0	1	#18	Medium		Coarse
1	.5	#35			Medium
2	.25	#60	Fine		Fine
3	.125	#120			Very Fine
4	.062	#230	Aphanitic Coarse	Silt	Coarse
5	.031	#400 (.037 mm)			Medium
6	.016		Fine		Fine
7	.008				Very Fine
8	.004			Clay	
9	.002		Microcrystalline to glass		
10	.001				

← The smallest particle visible with the naked eye (at 4 φ / .062 mm)

MINERAL HARDNESS

Mineral hardness is a relative hardness scale based off minerals given a "standard" hardness, usually in order from softest to hardest. Although a relative scale, it is very useful in identifying minerals when combined with other diagnostic tests. There are different hardness tests. The Vickers hardness test is very technical and not commonly used due to the expensive equipment involved. Most geologist use Mohs hardness scale.

Mohs Hardness Scale

The Mohs hardness scale is a good way to narrow down the possible range of your sample mineral. It is a scale of 1 to 10, with 1 being the softest and 10 being the hardest. We often use everyday items in the field to test mineral hardness. If the object scratches the mineral, then the object is harder. If the mineral scratches the object, then the mineral is harder than the object.

Below is a chart showing relative hardness of the defining minerals.

Defining Mineral	Scale Value	Common Object (Hardness)	Common Mineral (Hardness)
Diamond	10		
Corundum	9		
Topaz	8	Mason Drill Bit (8.5)	
Quartz	7		Pyrite (6-6.5)
Orthoclase	6	Steel Nail (6.5)	Olivine (6.5-7)
Apatite	5	Glass Plate/Knife (5.5)	Hornblende (5-6) Goethite (5-5.5)
Fluorite	4		
Calcite	3	Pre 1982 Penny (3.5)	Dolomite (3.5-4)
Gypsum	2	Fingernail (2.5)	Biotite (2.5-3) Muscovite (2-2.5)
Talc	1		
	Gypsum		

Common Igneous Minerals

Remember, when running a hardness test, you have to make sure that you are pressing into the mineral only. Do not cross over to other minerals. If crystals aren't large enough to be scraped, then you cannot perform a hardness test. When conducting the test, use a significant bit of pressure. Don't just lightly drag the tool over the mineral. That being said, don't press so hard that you are trying to dig in, you could break the tool. Below and on the next page, are charts of common mineral hardness in igneous rocks. **Bold** text = defining mineral. Red + **bold**, if defining mineral does *NOT* usually occur in igneous rock.

MINERAL	HARD-NESS
Albite (alkali feldspar)	6-6.5
Andesine (plagioclase)	6-6.5
Anorthite (plagioclase)	6-6.5
Anorthoclase (alkali feldspar)	6
Apatite	5
Augite	5.5-6
Beryl	7.5-8
Biotite	2.5-3
Bytownite (plagioclase)	6-6.5
Calcite	3
Chlorastrolite (Isle Royale Greenstone)	*See Pumpel-*
Chlorite	2-3
Copper (native element)	2.5-3
Corundum	9

MINERAL	HARDNESS RANGE
Datolite	5-5.5
Diamond	10
Dolomite	3.5-4
Epidote	6.5
Feldspars (general)	6-6.5
Fluorite	4
Garnet	6.5-7.5
Goethite	5-5.5
Gold (native element)	2.5-3
Gypsum	2
Hematite	5-6
Hornblende	5-6
Ice (0°C and 1 atm)	1.5

MINERAL	HARDNESS RANGE	MINERAL	HARDNESS RANGE
Ice (range)	1.5-6	**Quartz**	**7**
Iron (native element)	4-5	Sanidine (alkali feldspar)	6
Labradorite (plagioclase)	6-6.5	Silver (native element)	2.5-3
Lead (native element)	1.5	Siderite	4
Leucite (feldspathoid)	5.5-6	Sodalite (feldspathoid)	5.5-6
Magnetite	5.5-6	Sphene (a.k.a. Titanite, does *NOT* occur with Perovskite)	5-5.5
Malachite	3.5-4		
		Spinel	7.5-8
Mica (general)	1-6 (typically <3)	**Talc**	**1**
Microcline (alkali feldspar)	6-6.5	Tin	2
Molybdenite	1-1.5	Titanite	*See Sphene*
Muscovite	2-2.5		
		Topaz	**8**
Nepheline (feldspathoid)	5.5-6	Tremolite	5-6
Oligoclase (plagioclase)	6-6.5	Uraninite	5-6
Olivine	6.5-7	Zircon	6-7.5
Orthoclase (alkali feldspar)	**6**		
Perovskite (does *NOT* occur with Sphene a.k.a. Titanite)	5.5		
Platinum (native element)	4-4.5		
Pumpellyite (*see also Chlorastrolite*)	5.5		
Pyroxene Group	5-7		

UNDERSATURATED VS. SATURATED

Saturated and undersaturated are terms used herein to denote the amount of primary quartz (silica or SiO_2) that is in a rock. This only deals with primary quartz or the visible quartz in the rock that formed when the magma cooled. Although saturation is not key to describing a rock, nor is it necessary to put it in the description, it appears so much in the literature that it needs explaining.

Unsaturated rocks are the ones that contain very little to no primary quartz. This includes all foids on the QAPF diagram, since primary quartz cannot form with foids. It also includes all felsic rocks with <5% quartz. Ultramafic and mafic rocks are also generally unsaturated.

Undersaturated rocks contain little to some primary quartz $\geq$5% to 20%. Undersaturated rocks only fall on the QAP part of the QAPF diagram. Both plutonic and volcanic rocks can be undersaturated. Some mafic rocks can be undersaturated but they are rare in the Midwest, although some do exist in the greater Sudbury Ontario area.

Saturated rocks are ones with $\geq$20% to 60% primary quartz. Saturated rocks only fall on the QAP part of the QAPF diagram. Both plutonic and volcanic rocks can be saturated.

Oversaturated rocks contain $\geq$60% to 90% primary quartz and are pretty much restricted to plutonic rocks, like quartz granite. Although quartz rhyolites are volcanic, and fall in this range; they are not known to exist in the Midwest.

Supersaturated rocks are rocks $\geq$90% quartz. This is pretty much restricted to plutonic quartzolite. There appears to be some plutonic supersaturated igneous rocks within the Michigamme Formation at Horserace Rapids, and along US-141 in the Upper Peninsula of Michigan.

The QAPF diagrams on the next page highlight the areas showing the undersaturated, saturated, oversaturated, and supersaturated rock.

25

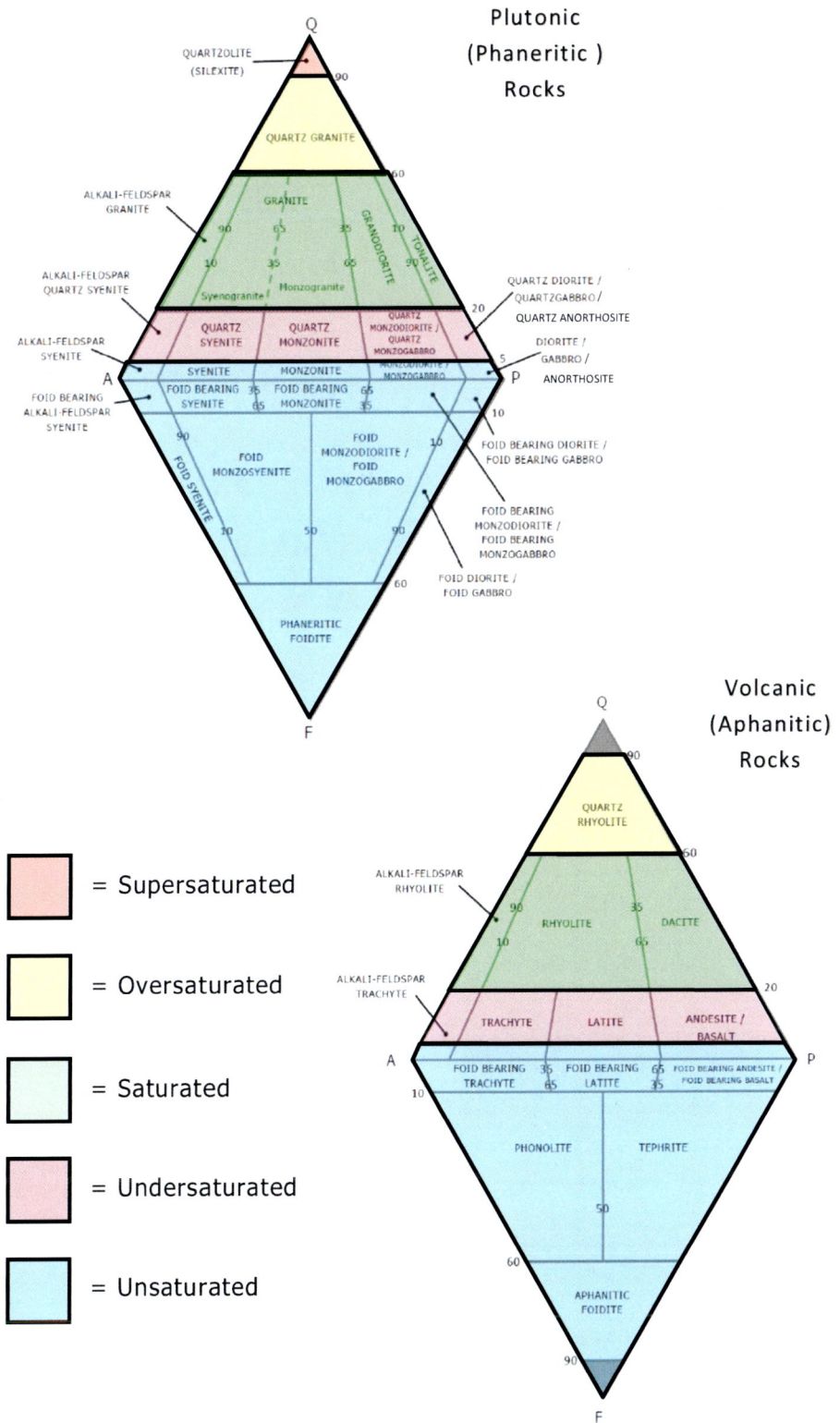

BOWEN'S REACTION SERIES

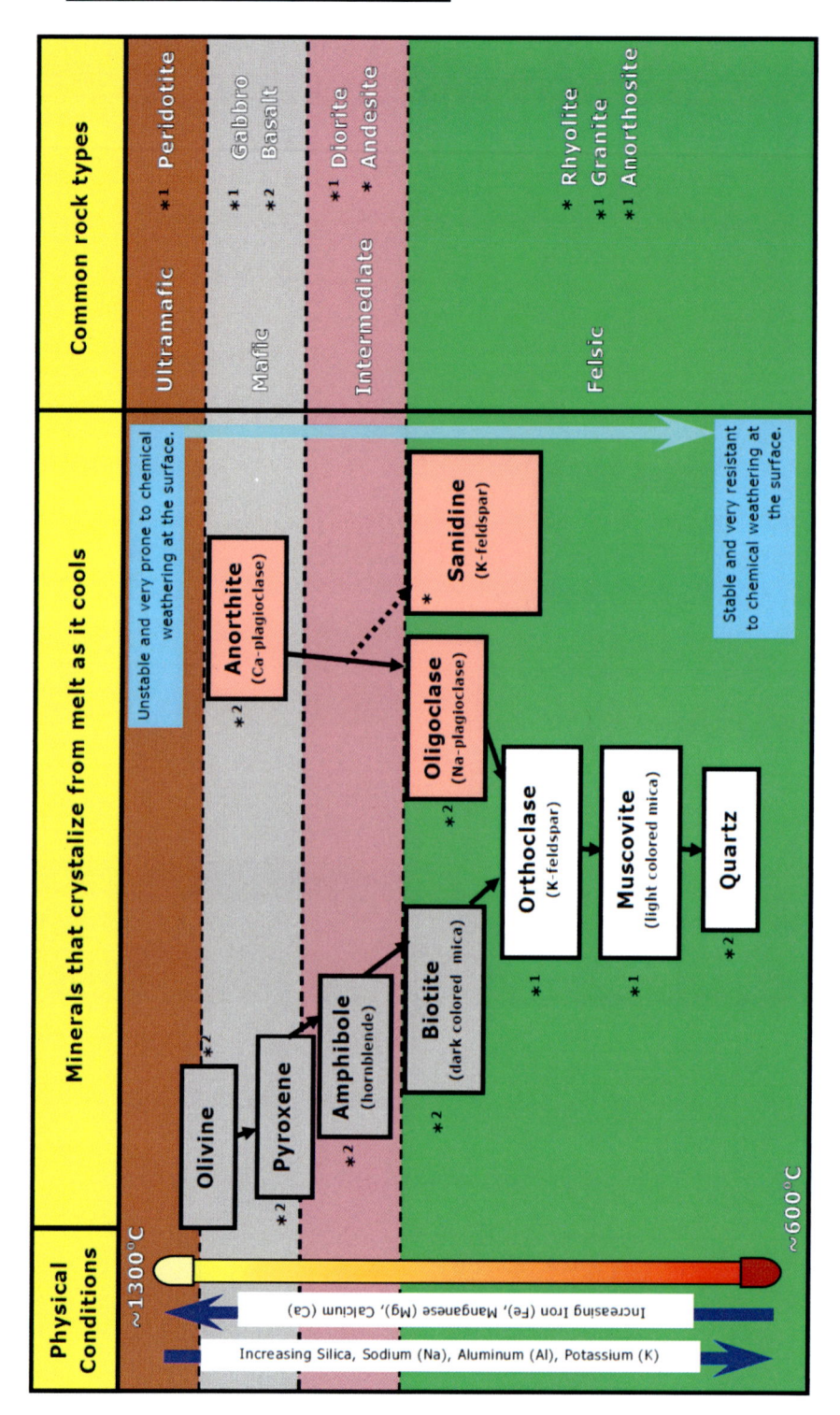

Adapted from Bowen 1922

Discontinuous series

If there is enough silica in the molten material, each mineral will change to the next lower mineral in the series as the temperature drops.

Continuous series

The continuous change in composition of a mineral in order to maintain a state of equilibrium within a molten material. The mineral changes its composition by continuously exchanging cations with the cooling molten material.

Residual phases

Once the molten temperature is lowered enough, and dependent of the water content and pressure, eventually only these three minerals will form in sequence from top to bottom.

* = Strictly volcanic in nature
*[1] = Strictly plutonic in nature
*[2] = Volcanic or plutonic in nature

NOTES: The rock anorthosite is *NOT* necessarily composed of the mineral anorthite. The two names being similar is a coincidence.

Sanidine is a high temperature K-Feldspar polymorph of orthoclase. Both have the same chemical structure (like microcline and orthoclase) but each have a different crystal structure.

Chart adapted from Bowen, 1922.

Bowen's Reaction Tree

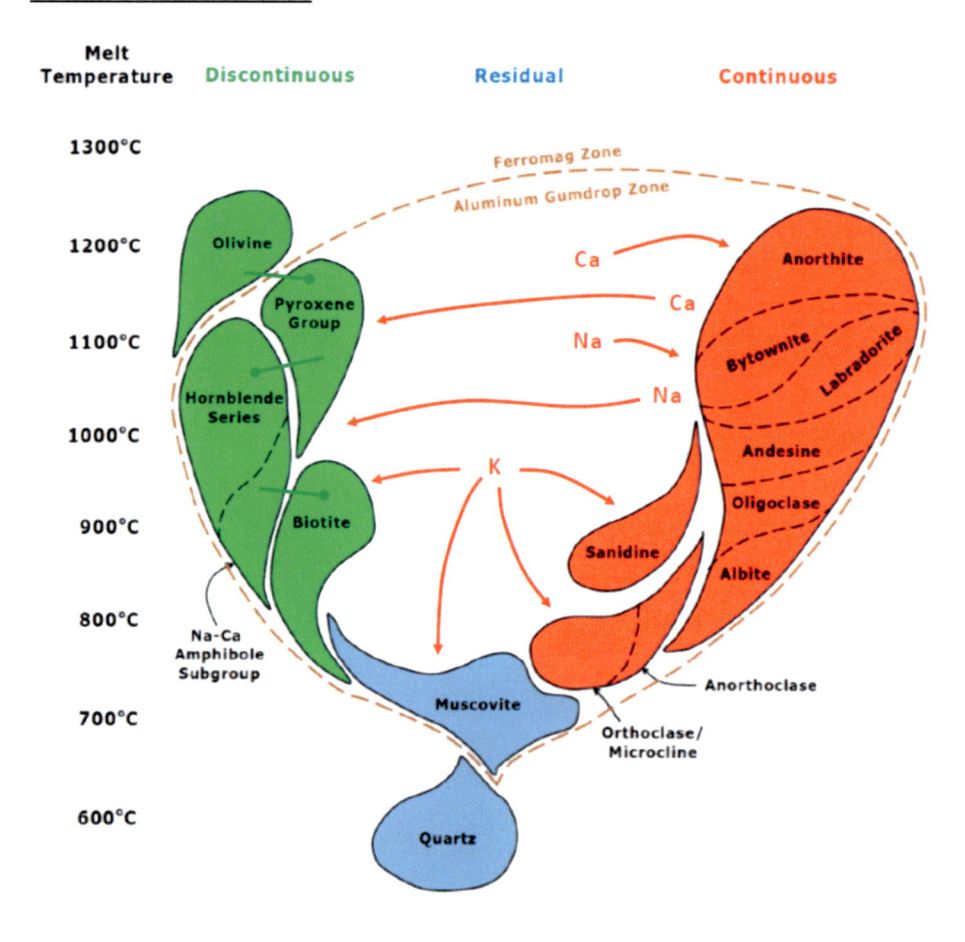

Melt Temperature — **Discontinuous** — **Residual** — **Continuous**

Temperature scale: 1300°C, 1200°C, 1100°C, 1000°C, 900°C, 800°C, 700°C, 600°C

Ferromag Zone
Aluminum Gumdrop Zone

Discontinuous series (green): Olivine, Pyroxene Group, Hornblende Series, Biotite, Na-Ca Amphibole Subgroup

Continuous series (orange): Anorthite, Bytownite, Labradorite, Andesine, Oligoclase, Albite, Sanidine, Anorthoclase

Residual (blue): Muscovite, Orthoclase/Microcline, Quartz

Arrows labeled: Ca, Ca, Na, Na, K

K → Elements that exchange ions with the continuous series and the melt and sometimes return to the continuous series but also get used by the discontinuous and residual series as the melt cools.

→● Showing how the discontinuous series molecules rearrange themselves into new minerals as the melt cools

This chart is my own derivation based on Bowen's reaction series. I created it to be more organic and fluid. Also I am attempting to highlight more of the gray areas in the series. Although melt temperature guides the mineralogy, chemistry determines the mineralogy.

The Aluminum (Al) Gum Drop Zone is where Al starts to crystalize out of the melt with other elements. The ferromag zone is the area where it's too hot for Al to crystallize out of the melt so only iron (Fe) and Magnesium (Mg) can crystallize out. Although Fe and Mg can crystallize out in the Al Gumdrop Zone as well. By the time you get to quartz, everything except silicon and oxygen has been used up.

At the dawn of the 20th century, Norman L. Bowen and others, began experimenting with rocks in order to see if certain minerals would crystalize from a magma first. Several things were discovered out of the experiments. First, the magma (or melt) would try to stay in equilibrium with the forming crystals (or minerals). This would result in different compositions between the melt and minerals, and would through off equilibrium. So the new crystals would re-react with the melt to form new minerals. Second, the newly crystalized minerals would form in a specific order. Third, if the magma had enough silica and was homogenous two main series would form. The continuous and discontinuous series, which would both merge to a simpler chemical melt composition of the magma as the iron, magnesium, calcium, sodium, etc. will be used up to form the residual series.

The continuous series deals with the crystallization of the feldspars. First plagioclase minerals will form. As the magma cools, this will throw off the equilibrium. So some of the calcium in the plagioclase will re-react with the magma and become potassium/sodium enriched until finally orthoclase crystalizes out.

The discontinuous series is odd, but makes sense. Say the magma produces olivine at a high temperature. As it cools further the olivine reacts with the melt but doesn't exchange ions like in the continuous series. Instead it will stop forming and pyroxene will form through a rearrangement of atoms.

The chemical reaction would look something like this (from a generic olivine to pyroxene): $Mg_2SiO_4 + SiO_2 \longrightarrow 2MgSiO_3$.

There is an overlap between when one mineral stops forming and another starts and you can get either or, or both. So, instead of say calcium being replaced with potassium, the increased silica and lower temperature causes the existing mineral to reorganize itself but made of the same elements. Basically all that is occurring is there is an internal crystal lattice (or structural) adjustment to achieve crystalline stability at lower temperatures until orthoclase is formed. Whether continuous or discontinuous, once orthoclase forms, only the residuals remain. Muscovite will not form before all the Fe and Mg in the melt is used up.

How does this help us identify minerals? Olivine and anorthite can form together and commonly do. But pyroxene and quartz cannot form at the same time since one pyroxene forms at a much higher temperature.

For example, say you have a fine to medium grained igneous rock and you see about 50% very dark but unidentifiable mineral, and 50% nearly white and heavily striated mineral. You can't identify the dark mineral but it can be assumed to be pyroxene. Why? Because if the other mineral is white and heavily striated, it has to be anorthite. Both are high temperature minerals in different series and are known to form together! Be careful. Just because anorthite and quartz can't form at the same time, that doesn't mean they can't be in the same rock. as quartz can still form at low magma temperatures if the magma is saturated in silica.

Bowen's reaction series doesn't account for everything. In Bowen's day, we didn't know about plate tectonics. The rolls of saturated verses unsaturated magma, incorporation of country rock, and the differentiation of magma due to cumulate, was only beginning to be studied. Bowen's reaction series is very useful but it deals with idealized conditions. This is why field mapping in conjunction with laboratory petrographic analysis is so important.

SIMPLIFIED FLOWCHART FOR CLASSIFYING IGNEOUS ROCKS

This is a basic flow chart showing you how to begin the process of classifying igneous rocks. Carbonate igneous rocks do not appear on this flowchart because igneous carbonates are practically non existent in the Midwest. This chart assumes all volcanic rocks can be plotted on the QAPF diagram (except for pyroclastics, lamprophyres, and ultramafics). If a rock is mafic (50-90% mafic minerals), you can use the mafic classification charts (p.35-37), but in most cases the QAPF diagrams are sufficient. Also, magnesium is not taken into account as it would require chemical analysis and eventually lead to a plot on a TAS diagram, which is beyond the scope of this book. The Melilite Group of minerals is also ignored as it is not prevalent in most igneous or metamorphic rocks of the Midwest. Pyroclastic rocks occur in the Proterozoic rocks, but some bear similarities to sedimentary rocks, thus warranting leaving them off this flowchart.

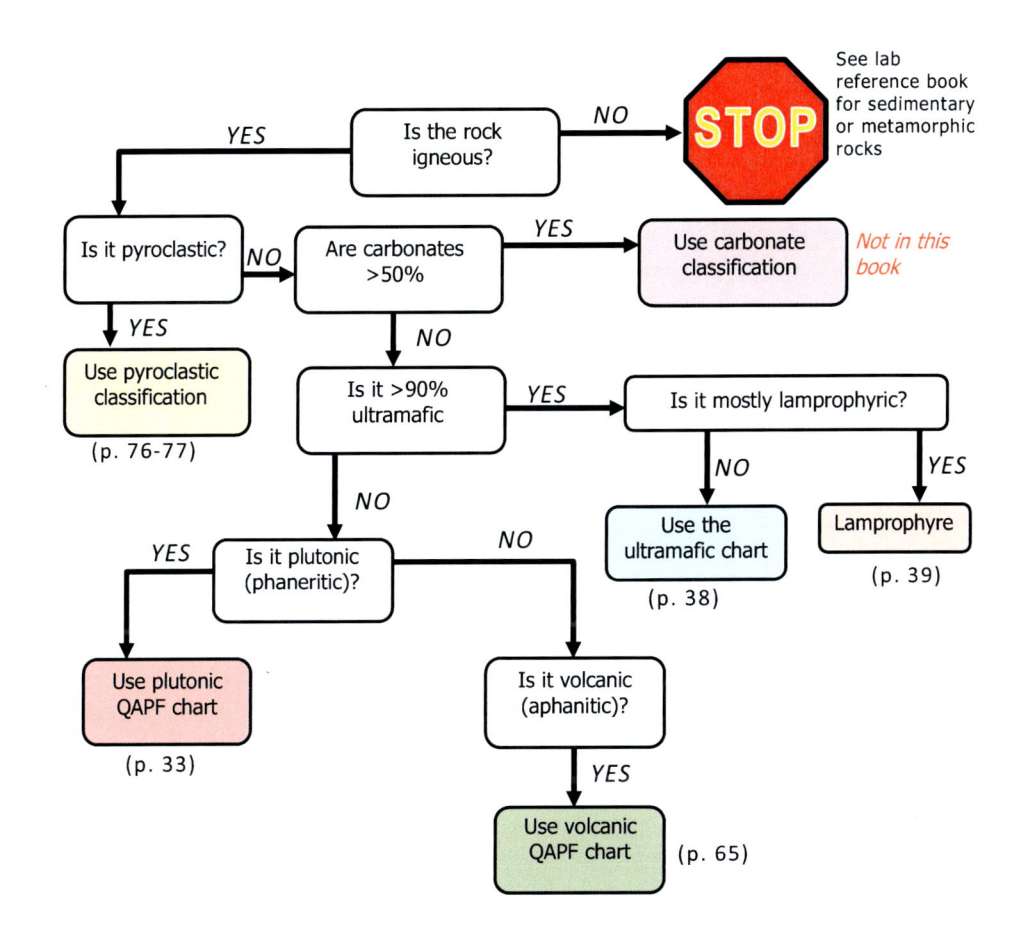

Adapted from: LeBas and Streckeisen (1991)

PART II:
CHARACTERISTICS OF PHANERITIC IGNEOUS ROCKS

IUGS QAPF PLOT FOR PHANERITIC (PLUTONIC) ROCKS

(2nd order normalization)

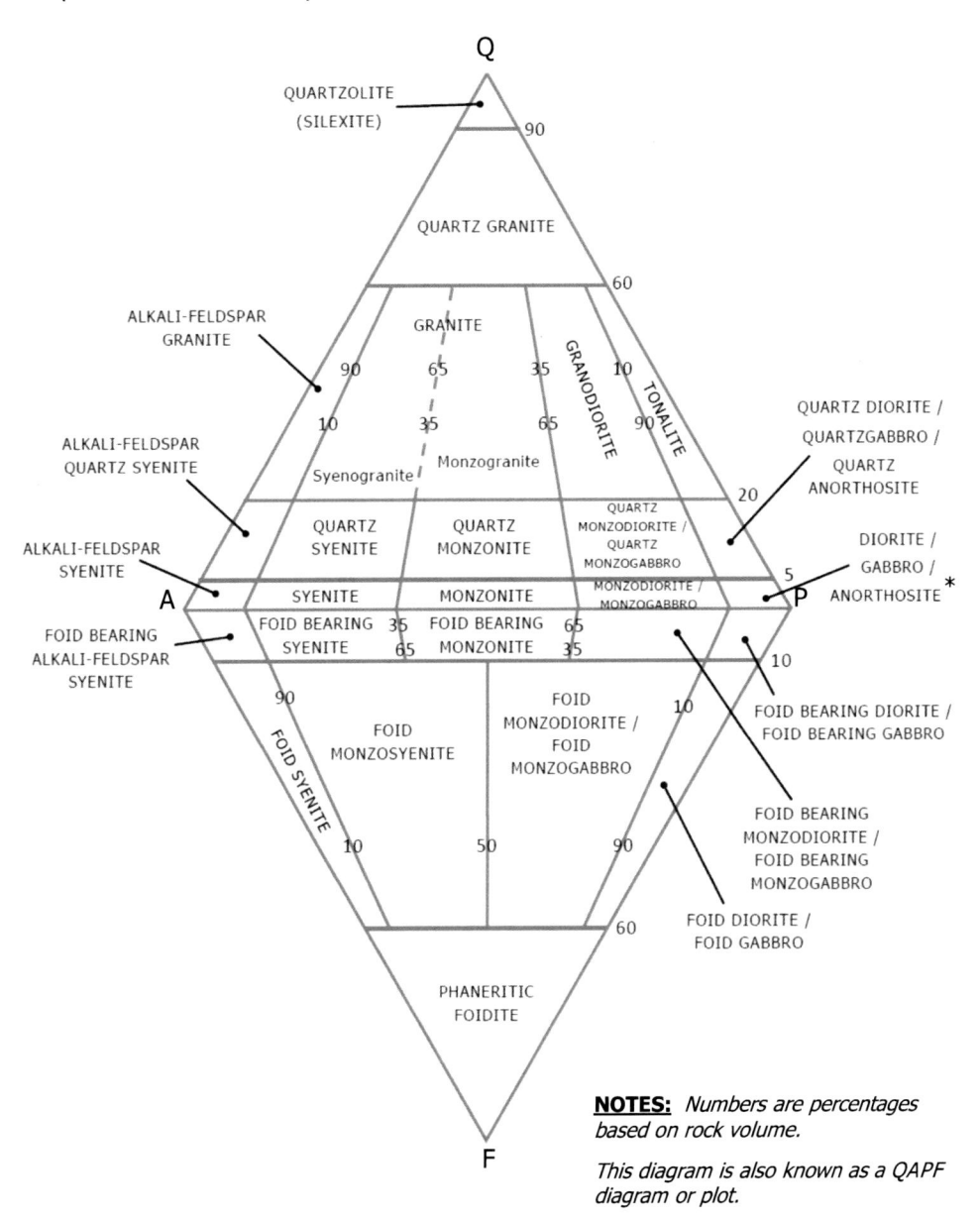

Q

QUARTZOLITE (SILEXITE)

90

QUARTZ GRANITE

60

ALKALI-FELDSPAR GRANITE

GRANITE

90 65 35 10

GRANODIORITE

TONALITE

10 35 65 90

QUARTZ DIORITE / QUARTZGABBRO / QUARTZ ANORTHOSITE

ALKALI-FELDSPAR QUARTZ SYENITE

Syenogranite Monzogranite

20

QUARTZ SYENITE

QUARTZ MONZONITE

QUARTZ MONZODIORITE / QUARTZ MONZOGABBRO

DIORITE / GABBRO / ANORTHOSITE *

ALKALI-FELDSPAR SYENITE

SYENITE

MONZONITE

MONZODIORITE / MONZOGABBRO

5

A

P

FOID BEARING ALKALI-FELDSPAR SYENITE

FOID BEARING SYENITE 35

FOID BEARING MONZONITE 65

65 35

10

90

FOID MONZODIORITE / FOID MONZOGABBRO

10

FOID BEARING DIORITE / FOID BEARING GABBRO

FOID SYENITE

FOID MONZOSYENITE

FOID BEARING MONZODIORITE / FOID BEARING MONZOGABBRO

10 50 90

FOID DIORITE / FOID GABBRO

60

PHANERITIC FOIDITE

F

NOTES: Numbers are percentages based on rock volume.

This diagram is also known as a QAPF diagram or plot.

* Anorthosite has NO volcanic equivalent.

Q = Quartz
A = Alkali feldspar
P = Plagioclase
F = Foids

Adapted from: LeBas and Streckeisen (1991), and Gill (2010)

The phaneritic rock classification is a ternary diagram is an adaptation from the International Union of Geological Sciences (IUGS) **iugs.org** adapted from Streckeisen and others, 2002.

This ternary plot is not like the traditional ternary plots used for unconsolidated deposits and sandstones (also known as the United States Department of Agriculture soil textural chart). This chart uses 3-points but you need to "normalize" (p. 67-69) all three first, and plot your quartz or your foids (p. 49-51). Just like you would for any ternary plot (p. 35,65).

REMEMBER: *Primary quartz and foids CANNOT crystalize from the same magma. They DO NOT occur in the same rock.*

From here, it differs. Then you plot your quartz or foids as a horizontal line. Once quartz or foids are plotted, then you must re-normalize your plagioclase and alkali feldspars before plotting them (p. 33,65). If you normalized them correctly they should be the same line.

Once the two lines are plotted, the point at which they meet, is your rock type.

This procedure works exactly the same for aphanitic (i.e. volcanic) rocks.

MAFIC AND ULTRMAFIC PHANERITIC (PLUTONIC) ROCK CLASSIFICATION

These plots are somewhat optional in the Midwest. I am personally not a fan of the mafic and ultramafic plots. It over complicates things in my eyes. However, since things like norite are known to occur in the Midwest (especially around Sudbury, Ontario); I felt they should be included.

The mafic and ultramafic diagrams on the next page function like the QAPF diagrams. Olivine and anorthite *CAN* occur together.

There is norite in Ontario, so you may find it useful. Mafic rocks contain about >50-90% mafic minerals. The 50% part can be a discretionary call if the rock displays larger feldspar crystals than mafic crystals or vise versa, but it shouldn't exceed 50±5%.

Ultramafic rocks contain >90% mafic minerals. The Mafic phaneritic dikes are very common in the Archean and Proterozoic plutonic felsic (rocks rich in feldspar and quartz, usually light colored) rocks in the Midwest. Common ultramafic rocks are peridotite, dunite (a.k.a. olivinite), pyroxenite, and rocks composed of the serpentine group of minerals (altered forms of olivine, usually through very low grade metamorphism). Although technically metamorphic, a serpentine rich ultramafic rock would be the altered form of dunite, it is often called "serpentinite".

PHANERITIC MAFIC

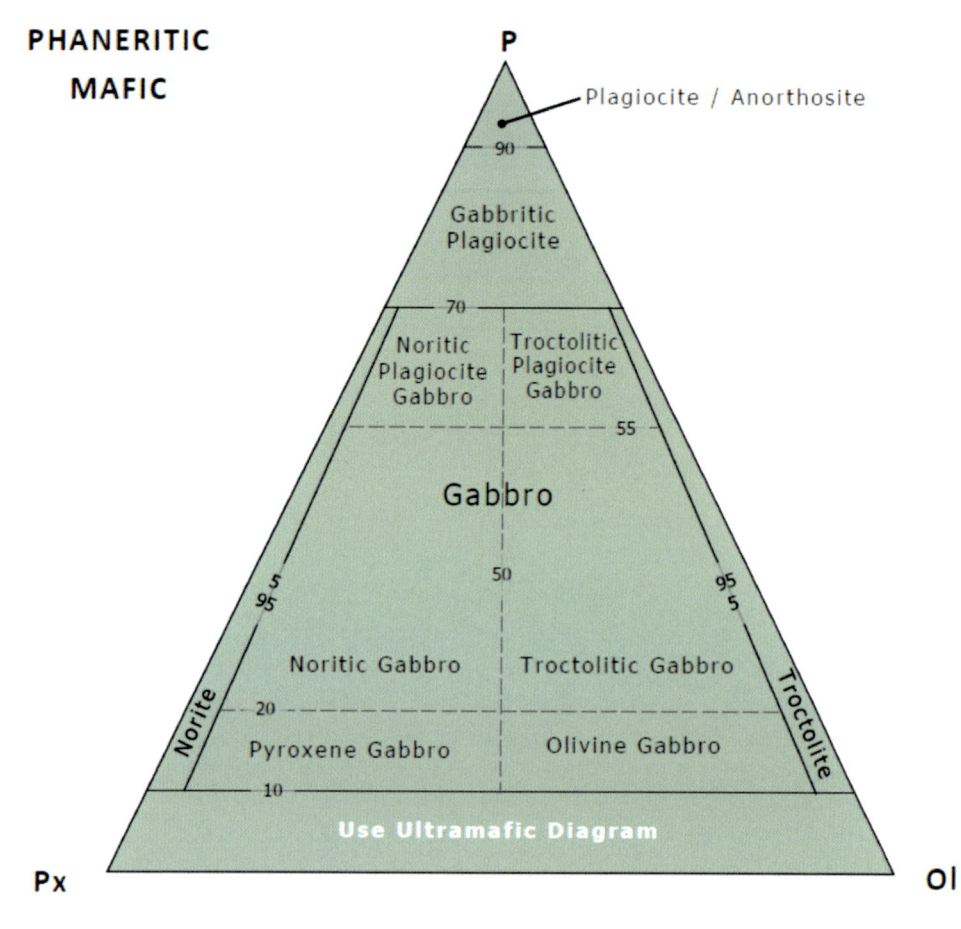

```
                    P
                    ▲
                   /90\  ← Plagiocite / Anorthosite
                  /     \
                 / Gabbritic \
                /  Plagiocite \
               /-------70------\
              / Noritic | Troctolitic \
             / Plagiocite| Plagiocite \
            /  Gabbro   |  Gabbro     \
           /------------|------55------\
          /                             \
         /            Gabbro             \
        /              50                 \
       /5                               95 \
       95                                 5
      /   Noritic Gabbro | Troctolitic Gabbro\
     /--------20---------|--------------------\
    /  Pyroxene Gabbro   |  Olivine Gabbro     \
   /----------10---------|----------------------\
  /        Use Ultramafic Diagram                \
 Px ──────────────────────────────────────────── Ol
```

(Norite along left edge; Troctolite along right edge)

Phaneritic mafic rocks:

Gabbros contain <5% quartz and <10% K-spar. If they are present, they need to be removed.

The plagioclase is >50% Anorthite, Bytownite, and Labradorite.

(2nd order normalization)

Adapted from: LeBas and Streckeisen (1991), Ciccolella (2016), and Dohaney, no date.

PHANERITIC MAFIC

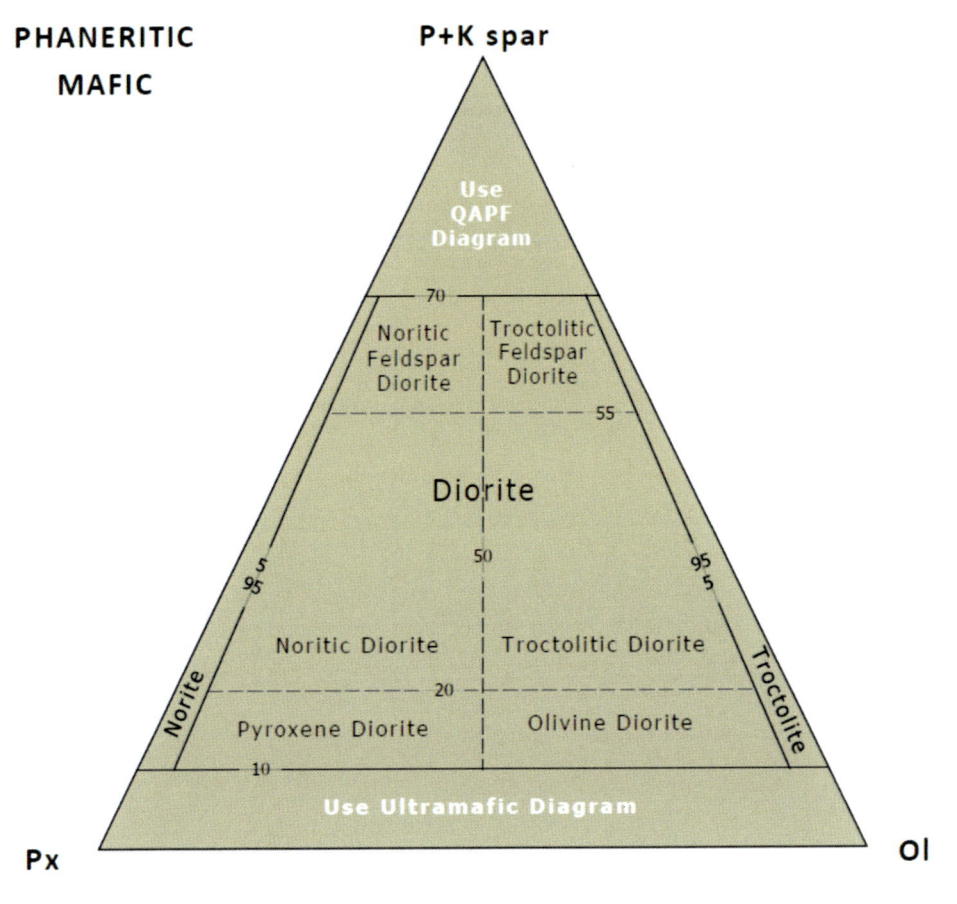

Triangular diagram with apices: P+K spar (top), Px (bottom left), Ol (bottom right).

Regions within the triangle:
- Use QAPF Diagram (top, above 70)
- Noritic Feldspar Diorite | Troctolitic Feldspar Diorite (between 70 and 55)
- Diorite (center, at 50)
- Noritic Diorite | Troctolitic Diorite (between 50 and 20)
- Pyroxene Diorite | Olivine Diorite (between 20 and 10)
- Use Ultramafic Diagram (bottom, below 10)
- Norite (left edge, 5/95)
- Troctolite (right edge, 95/5)

Phaneritic mafic rocks:

Diorites contain <5% quartz and <10% K-spar. If they are present, they need to be removed.

The plagioclase is >50% Andesine, oligoclase, and albite.

(2nd order normalization)

Streckeisen, A. (1974) Classification and Nomenclature of Plutonic Rocks. Geologische Rundschau, 63, 773-786.
https://doi.org/10.1007/BF01820841

PHANERITIC ULTRAMAFIC

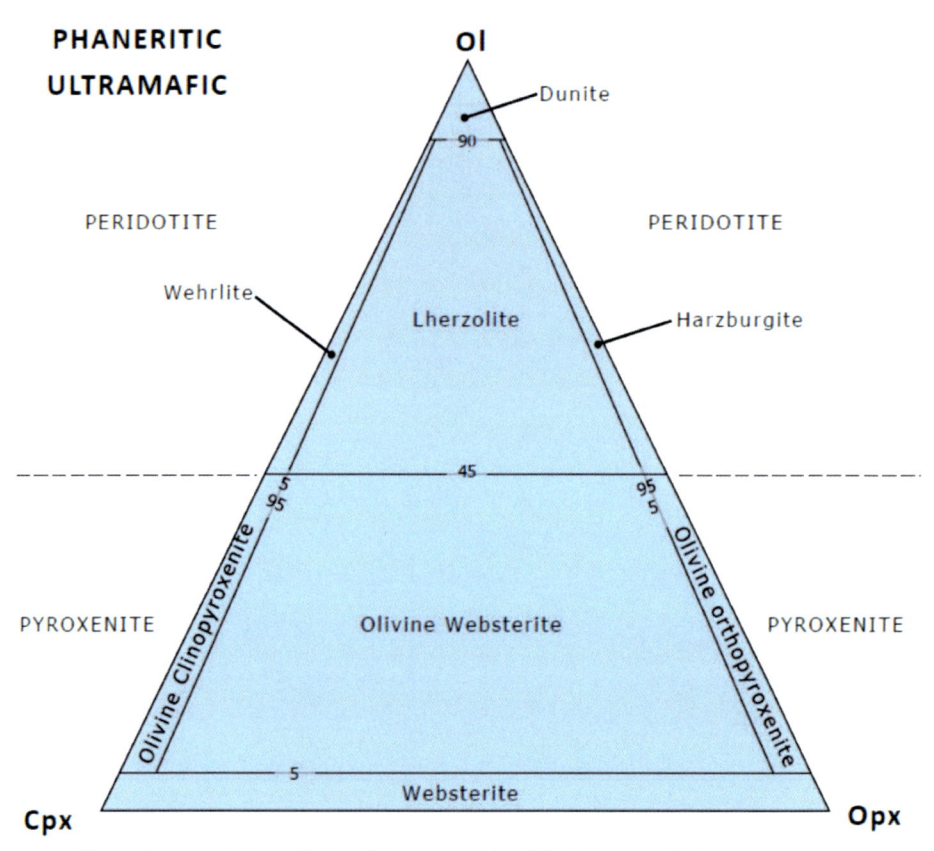

Ternary diagram with apices Ol (top), Cpx (bottom left), Opx (bottom right):
- Dunite (Ol >90%)
- PERIDOTITE field (Ol 45–90%): Wehrlite (Ol-Cpx side), Lherzolite (center), Harzburgite (Ol-Opx side)
- PYROXENITE field (Ol <45%): Olivine Clinopyroxenite, Olivine Websterite, Olivine orthopyroxenite, Websterite
- Boundary values shown: 90, 45, 5 (Ol); 5/95 on side boundaries

Ultramafic rocks (phaneritic): <5% quartz and <10% feldspars. If they are present, they need to be removed.

Without thin sections, it can be extremely difficult to differentiate Cpx from Opx. In that case a simplified plot can be used.

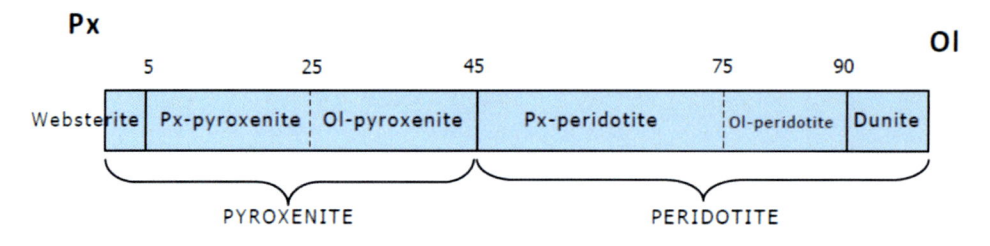

Linear plot from Px (left) to Ol (right) with divisions at 5, 25, 45, 75, 90:
- Websterite (0–5)
- Px-pyroxenite (5–25)
- Ol-pyroxenite (25–45)
- Px-peridotite (45–75)
- Ol-peridotite (75–90)
- Dunite (90–100)

PYROXENITE (0–45) | PERIDOTITE (45–100)

(2nd order normalization)

Unique Mafic and Ultramafic Rocks

Below is a chart for the general standard for lamprophyre classification. It is not perfect and some workers don't like it. But it is very usable and fits the criteria listed below. If needed for specialty work, lamprophyres can be divided in another way like granites are with the *SIAM classification. But keep the specialty classifications for targeted lab based papers.

For a classification system to be practical with as little specialty equipment as possible, it needs to be:

1) Simple and logical
2) Focused on the rocks targeted. Not tied into a bunch of related rocks.
3) Usable to a very large extent in the field
4) Lithologically and/or mineralogically based. No interpretive aspects.
5) Unique terms where possible to avoid confusion with other classification systems.

I use the below chart but I do have a problem with it. I see the point in separating foids from feldspars (although that can be hard to determine in the field). But lamprophyres, by definition, contain only a very small fraction of those minerals. But on the plus side, those amounts aren't needed for the three basic names in yellow.

An important note about lamprophyres in general. The should never contain quartz because they are very under saturated rocks. Lamprophyres are as about as ultramafic as a rock can be. If your lamprophyre contains foids it cannot contain quartz, because quartz and foids never form together.

Classification of Lamprophyres

Felsic minerals		Predominant Mafic Minerals			
		Calc-alkaline Lamprophyre		Alkaline Lamprophyre	Melilitic Lamprophyre
		biotite	hornblende	amphibole (barkevikite, Kaersuite)	melilite, biotite
		diopsidic augite	diopsidic augite	Ti-augite	±Ti-augite
Feldspar	Feldspathoid	(±olivine)	(±olivine)	olivine, biotite	±olivine
					±calcite
K-spar > plagioclase		minette	vogesite		
K-spar < plagioclase		kersanite	spessartite		
K-spar > plagioclase	feldspar > foid			sannaite	
K-spar < plagioclase	feldspar > foid			camptonite	
	glass or foid			monchiquite	polzenite
					alöite

▢ = rock name

Adapted from: LeBas and Streckeisen, 1991

This chart is by no means perfect but it is simple, useable in the field, and completely based off mineralogy. It is also the one accepted by the International Union of Geological Sciences (IUGS).

*SIAM stands for S-type granite, I-type granite, A-type granite, M-type granite. The system is a tectonic interpretation classification scheme and so it is beyond the scope of this book.

The chart below is not reliable in my opinion and I do not endorse its usage. I am only including it because it pops up online a lot.

I am not blasting the authors. I see what they were trying to do here. But as a general classification chart it is not practical, I didn't even include Krmicek and Rao's complete version of the chart (see below). It also breaks down kimberlites and lamproites which I believe is completely unnecessary for general field and lab use. So I left them out of the chart below.

1) The below chart is logical, but it is not simple. It creates a Linnaean style hierarchy by using branch, clan, and family. Which isn't necessarily bad, but it needs to be applied to all rocks if used. Not just lamprophyres.

2) The chart tries to tie in kimberlites and lamproites. Kimberlites are distinct enough, rare enough, and have economic value, so tying them into lamprophyres is not really necessary. I have personally never liked the term lamproite as it crosses many other rock types which are better defined, and is somewhat interpretive. The chart below is also interpretive under calc-alkaline lamprophyres, as the terms volcanic, plutonic, and hypabyssal are used.

3) The chart is only somewhat useable in the field, but not impossible to use.

4) It is lithologically based, up until the interpretive part comes in.

5) I feel the Linnaean style hierarchy could have been filled with different terms, as the ones presented are too similar to what is used in biology. I have the same issue the usage of series and suite, except more so. Series and suite hit closer to home than clan, branch, and family. Series and suite are used for two different classification systems in stratigraphy. A suite is used in lithostratigraphy under lithodemic classification, and series is used in chronostratigraphy. Their usage here, could cause confusion.

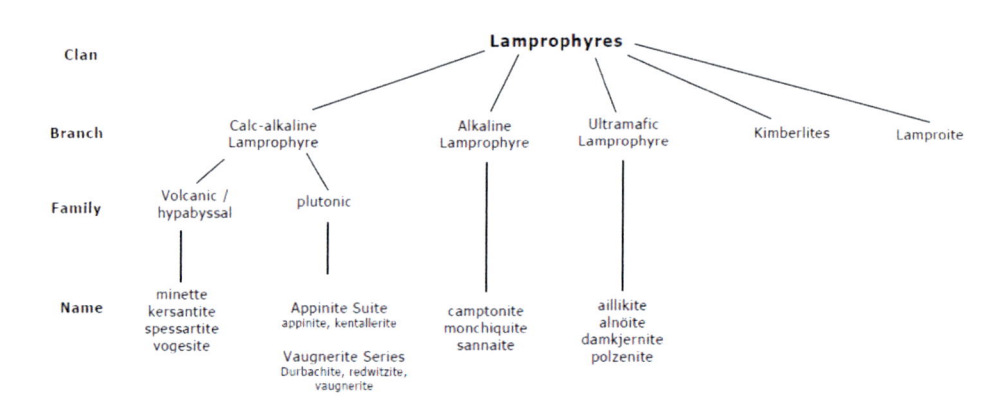

Clan			Lamprophyres			
Branch	Calc-alkaline Lamprophyre		Alkaline Lamprophyre	Ultramafic Lamprophyre	Kimberlites	Lamproite
Family	Volcanic / hypabyssal	plutonic				
Name	minette kersantite spessartite vogesite	Appinite Suite appinite, kentallerite Vaugnerite Series Durbachite, redwitzite, vaugnerite	camptonite monchiquite sannaite	aillikite alnöite damkjernite polzenite		

Adapted from: Krmicek and Rao, 2022

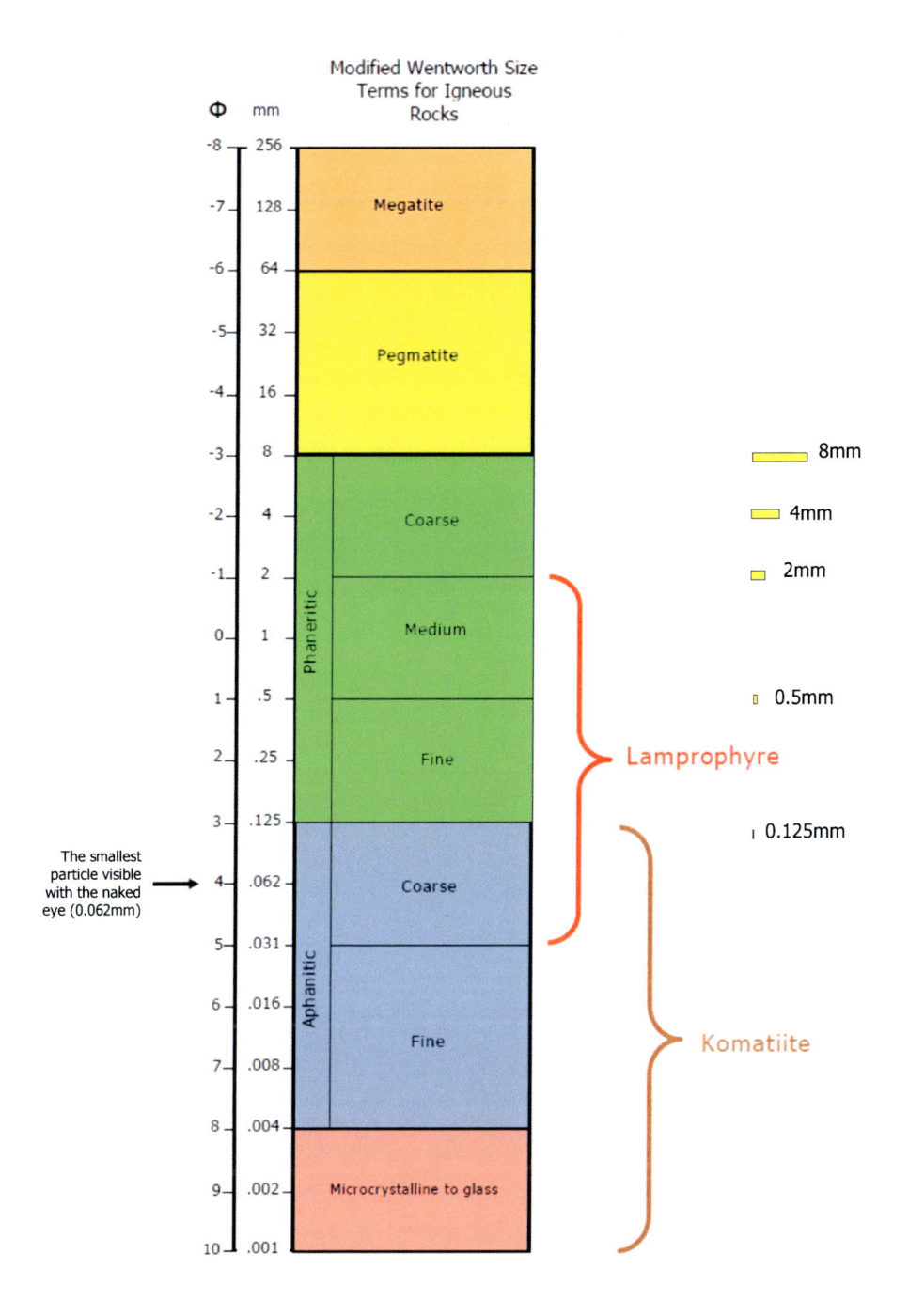

Modified Wentworth Size Terms for Igneous Rocks

Φ	mm		
-8	256	Megatite	
-7	128		
-6	64	Pegmatite	
-5	32		
-4	16		
-3	8	Coarse	Phaneritic
-2	4		
-1	2	Medium	
0	1		
1	.5	Fine	
2	.25		
3	.125		
4	.062	Coarse	Aphanitic
5	.031		
6	.016	Fine	
7	.008		
8	.004		
9	.002	Microcrystalline to glass	
10	.001		

The smallest particle visible with the naked eye (0.062mm)

- 8mm
- 4mm
- 2mm
- 0.5mm
- 0.125mm

Lamprophyre (Phaneritic fine through Aphanitic coarse)

Komatiite (Aphanitic coarse through Microcrystalline to glass)

41

PLAGIOCLASE VS. ALKALI FELDSPARS (K-SPAR)

Perhaps one of the key things to recognize in an igneous rock is the difference between plagioclase and alkali feldspars. Before I get into physical ways these two differ, I need to exactly what these things exactly are.

Assuming all of the chemicals in a magma body are present in relative abundance, and no new material is being added or taken away, the magma (or lava) as it cools, will follow Bowen's Reaction series (p. 26-28). The plagioclase to alkali feldspar reactions are on the continuous side of the reaction. That means as the magma cools calcium ions will be replaced with sodium, and eventually potassium, before a solid state is reached. It does this by exchanging cations with the molten magma as it cools. The compositional phase diagram below shows how the exchange progresses from An to Or.

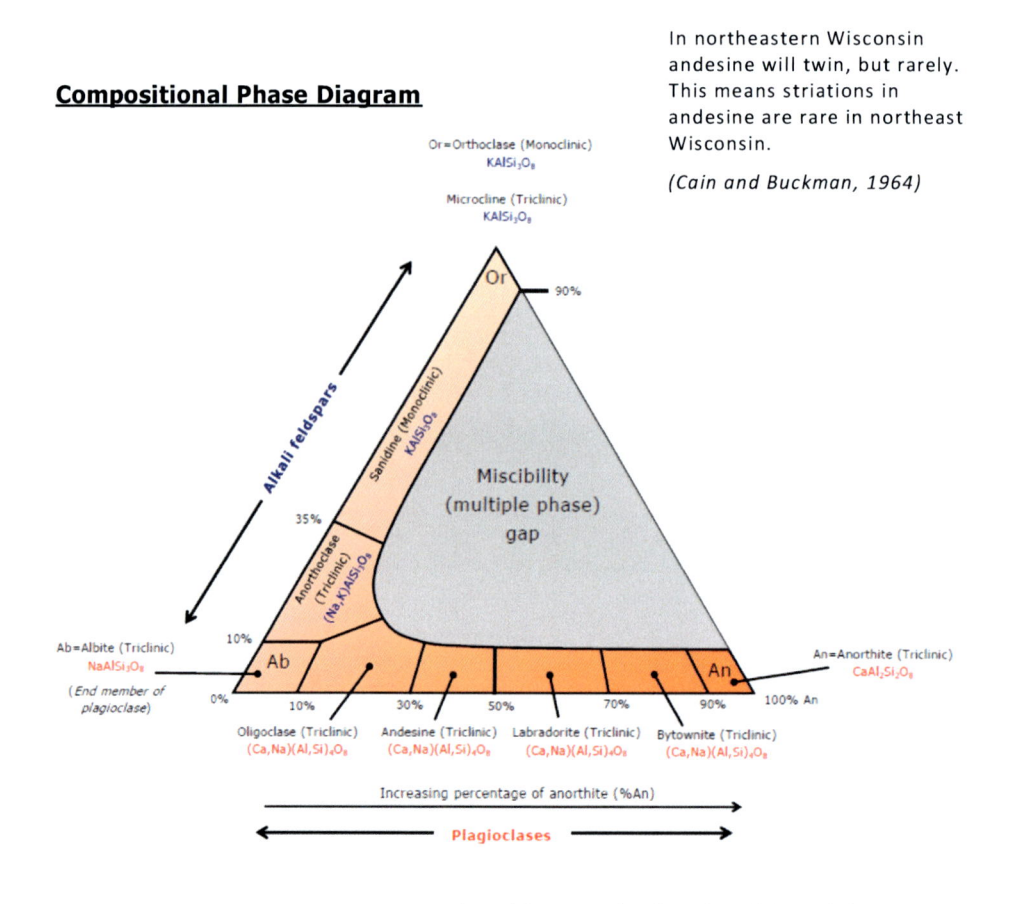

Compositional Phase Diagram

In northeastern Wisconsin andesine will twin, but rarely. This means striations in andesine are rare in northeast Wisconsin.

(Cain and Buckman, 1964)

Or = Orthoclase (Monoclinic) $KAlSi_3O_8$
Microcline (Triclinic) $KAlSi_3O_8$
Sanidine (Monoclinic) $KAlSi_3O_8$
Anorthoclase (Triclinic) $(Na,K)AlSi_3O_8$

Miscibility (multiple phase) gap

Ab = Albite (Triclinic) $NaAlSi_3O_8$
(End member of plagioclase)

An = Anorthite (Triclinic) $CaAl_2Si_2O_8$

Oligoclase (Triclinic) $(Ca,Na)(Al,Si)_4O_8$
Andesine (Triclinic) $(Ca,Na)(Al,Si)_4O_8$
Labradorite (Triclinic) $(Ca,Na)(Al,Si)_4O_8$
Bytownite (Triclinic) $(Ca,Na)(Al,Si)_4O_8$

Increasing percentage of anorthite (%An)

← Plagioclases →

Adapted from: Gordon (2015), Fedele et al. (2015)

Striations on Plagioclase vs. Microcline

Compositional phase diagram shows the path the minerals develop as the melt (magma or lava) cools, it only slightly contributes to identifying the feldspars. By the diagram we can tell that all plagioclase is triclinic, and alkali feldspars are usually monoclinic (except microcline).

Well developed crystals of feldspars are rare in igneous rocks. With one exception, there is a sure fire way to differentiate plagioclase from alkali feldspar. The presence of continuous parallel striations in a single direction on plagioclase. Be careful. Microcline is the *ONLY* alkali feldspar that will show striations. However, its striations are cross hatched and at nearly 90° to one another. The microcline striations are usually discontinuous. Orthoclase, unlike microcline doesn't exhibit striations.

Striations on Plagioclase

Striations on Microcline

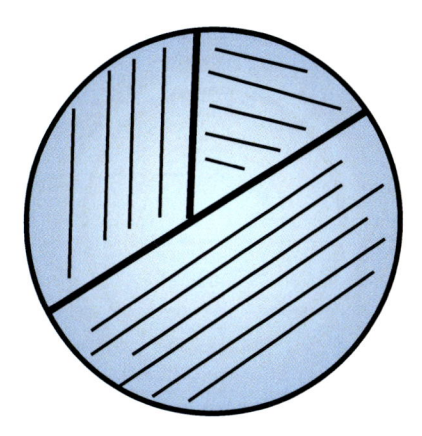

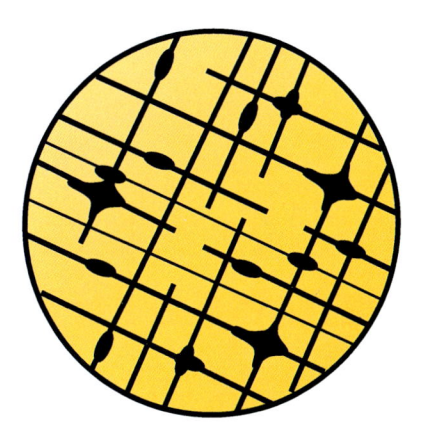

The thick black lines represent the boundaries between three different crystals in different orientations. The thin lines show how plagioclase striations can appear under magnification.

The striations on a single crystal of microcline as they can appear under magnification. Notice how they are discontinuous, cross hatched, and exhibit minor weathering at some junctions (black splotches).

Quick Flowchart for Choosing Plagioclase or Alkali Feldspar

This section is just meant to be a quick identification method. Do not use it exclusively without checking it against other measures. Plagioclase and alkali feldspars can be a wide range of colors. Color by itself is *NOT* diagnostic. Remember to look for striations and other diagnostic features.

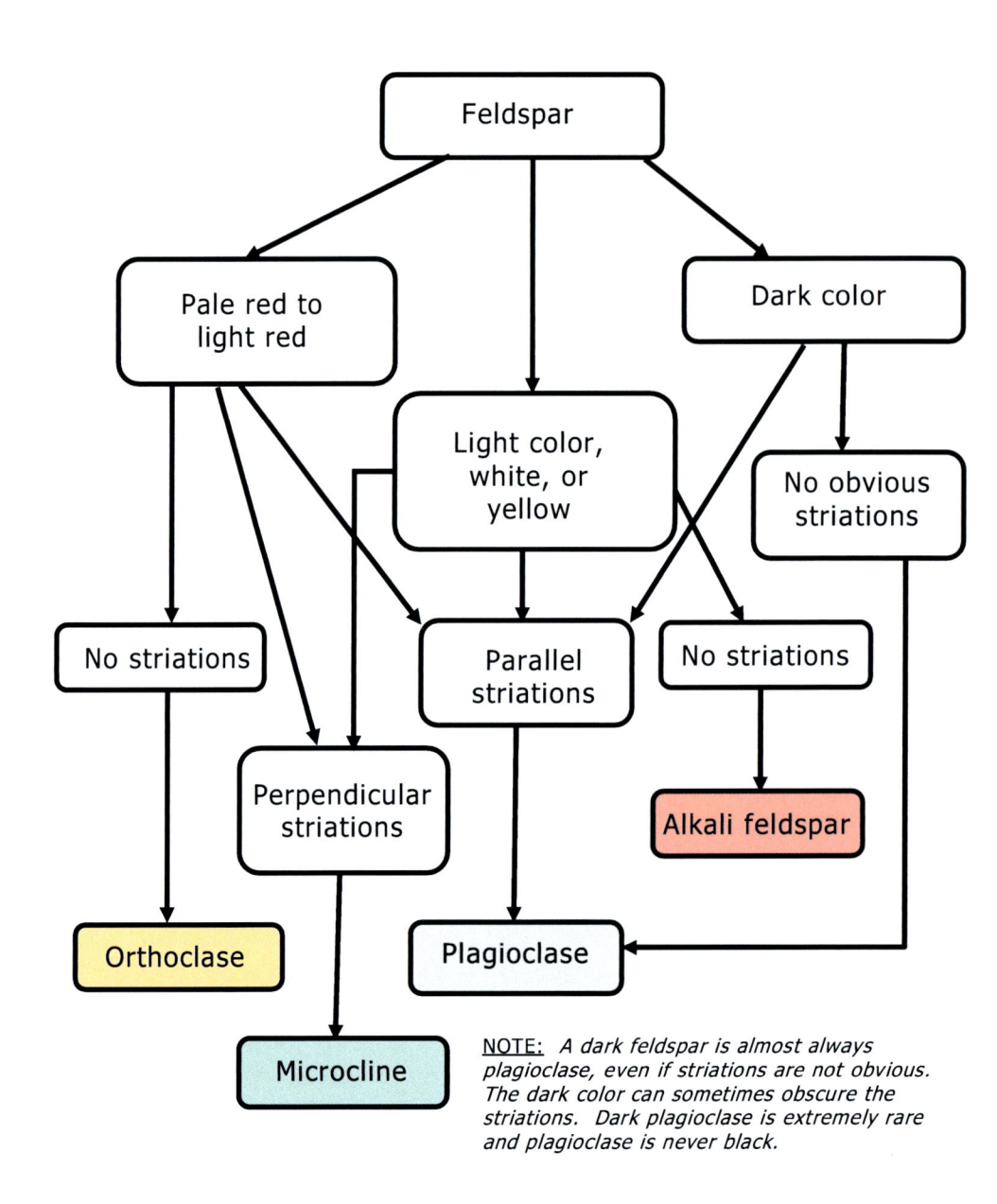

```
                           Feldspar
              /               |               \
    Pale red to        Light color,         Dark color
    light red          white, or            /        \
     /    |    \       yellow              /     No obvious
    /     |     \       |                 /       striations
   /      |      \      |                /           |
No        |    Perpendicular   Parallel   No striations
striations|    striations      striations       |
   |      |        |              |             |
Orthoclase|        |          Plagioclase ← ← ← Alkali feldspar
          |    Microcline
```

NOTE: A dark feldspar is almost always plagioclase, even if striations are not obvious. The dark color can sometimes obscure the striations. Dark plagioclase is extremely rare and plagioclase is never black.

OTHER KEY FACTORS DIFFERENTIATING ALKALI FELDSPAR FROM PLAGIOCLASE

Individual minerals can be very difficult to pick out without an electron microscope or good thin sections. Microcline and orthoclase can be identified with confidence (see previous two pages), but individual plagioclase really cannot. The reason for this is they all have nearly identical crystal structures as seen with the naked eye and even under magnification. They basically just contain varying amounts of sodium and calcium. Here are some basic mineral characteristics to use in order to tell them apart without thin section analysis. As you will see, the plagioclase series essentially has all the same properties except for common associated minerals. Labradorite is somewhat unique, due to its iridescent properties.

Albite

Chemical Formula: $NaAlSi_3O_8$ An<10%   Crystal System: Triclinic

Crystal Class: Pinacoidal

Color: White to gray, rarely with blue to green hues or red hues   Streak: White

Relative opacity: Transparent to translucent

Luster: Vitreous   Mohs Hardness: 6-6.5

Specific gravity: 2.55-2.65, average = 2.60

Fracture: Uneven to conchoidal   Habit: Massive, granular, tabular

Molecular mass: 263.02 g

Common associated minerals: microcline, quartz, hornblende

Other Characteristics:

Fluoresces weakly as a very deep red in long and shortwave UV, non magnetic, non radioactive. It exhibits deep parallel striations on good crystal faces. Melting point is 1100-1120°C.

It will occur with microcline in alternating patterns.

Oligoclase

Chemical Formula: (Ca,Na)(Al,Si)$_4$O$_8$: An=10-30%   Crystal System: Triclinic

Crystal Class: Pinacoidal

Color: White, sometimes gray, slight yellow brown and red hues, rarely green or blue

Streak: White

Relative opacity: Translucent to opaque

Luster: Vitreous   Mohs Hardness: 6-6.5

Specific gravity: 2.63-2.66, average = 2.65

Fracture: Irregular, uneven, subconchoidal   Habit: Massive, granular, tabular

Molecular mass: ~265 g

Common associated minerals: quartz, epidote, biotite

Other Characteristics:

Fluoresces light red in shortwave (if closer to albite in composition), non magnetic, non radioactive. It exhibits decent parallel striations on good crystal faces. Melting point is 1050-1150°C.

Andesine

Chemical Formula: (Ca,Na)(Al,Si)$_4$O$_8$: An=30-50%   Crystal System: Triclinic

Crystal Class: Pinacoidal

Color: White, sometimes gray, rarely slight yellow greenish or red orange

Streak: White

Relative opacity: Transparent to semi opaque

Luster: Semi vitreous to pearly   Mohs Hardness: 6-6.5

Specific gravity: 2.66-2.68, average = 2.67

Fracture: Irregular, uneven to conchoidal   Habit: Massive, granular, cleaves easily

Molecular mass: 268.62 g

Common associated minerals: quartz, magnetite, biotite, hornblende

Other Characteristics:

Fluoresces violet in medium to long wave UV (specifically 295-335 nm), non magnetic, non radioactive. It exhibits decent parallel striations on good crystal faces. Melting point is ~1150°C. Tends to be brittle, almost like mica. Does not usually make striations when twinning in the plutons of northeast Wisconsin. Rare in granitic rocks and syenites but common in andesites.

Labradorite

Chemical Formula: (Ca,Na)(Al,Si)$_4$O$_8$: An=50-70%   Crystal System: Triclinic

Crystal Class: Pinacoidal

Color: Gray, blue, occasionally greenish to light brown or colorless

Streak: White

Relative opacity: Transparent to translucent, always iridescent

Luster: Vitreous to pearly   Mohs Hardness: 6-6.5

Specific gravity: 2.68-2.72, average = 2.70

Fracture: Uneven to conchoidal   Habit: Massive, thin and tabular, rhombic appearing

Molecular mass: 271.81 g

Common associated minerals: Olivine, magnetite, pyroxenes, hornblende

Other Characteristics:

Rarely fluoresces but will sometimes fluoresce in purple (longwave) and red (shortwave) with certain activators, non magnetic, non radioactive. It exhibits deep and well defined parallel striations on good crystal faces. Melting point is ~1200°C.

The iridescent properties of labradorite make it appear to be different colors that "move" when looked at from different angles. The iridescence can make bluish to a full rainbow of colors (depending on the impurities in the mineral).

Bytownite

Chemical Formula: (Ca,Na)(Al,Si)$_4$O$_8$: An=70-90%   Crystal System: Triclinic

Crystal Class: Pinacoidal

Color: White, colorless, hues of brown to yellow brown

Streak: White

Relative opacity: Transparent to translucent

Luster: Vitreous to pearly on cleavages   Mohs Hardness: 6-6.5

Specific gravity: 2.72-2.74, average = 2.73

Fracture: Uneven to conchoidal   Habit: Massive, rarely crystalline, brittle

Molecular mass: 271.81 g

Common associated minerals: Olivine, pyroxenes

Other Characteristics:

Non fluorescent, non magnetic, non radioactive. It exhibits poorly to moderately defined parallel striations on good crystal faces. It can be iridescent if closer to labradorite in composition. Melting point is ~1250°C.

Anorthite

Chemical Formula: $(Ca,Na)(Al,Si)_4O_8$: An>90%   Crystal System: Triclinic

Crystal Class: Pinacoidal

Color: White, gray, occasionally light red to red hues

Streak: White

Relative opacity: Transparent to semi opaque

Luster: Vitreous   Mohs Hardness: 6-6.5

Specific gravity: 2.72-2.75, average = 2.74

Fracture: Uneven to conchoidal   Habit: Granular, anhedral-subhedral

Molecular mass: 277.41 g   Formula mass: 278.203g·mol^{-1}

Common associated minerals: Olivine, pyroxenes, corundum

Other Characteristics:

Fluoresces yellow-white (if corundum is present) in shortwave, non magnetic, non radioactive. It exhibits moderately to well defined parallel striations on good crystal faces. Melting point is ~1300°C.

All fluorescent data on p.45-48, was taken from the Franklin Museum's website.

FELDSPATHOIDS (FOIDS)

In the Midwest, there are only about three foids you will likely run across. They are nepheline, sodalite, and leucite. Most of these minerals are in place in the Colwell Complex of Ontario, between Marathon and Terrace Bay. Remember, you cannot have foids in a rock with primary quartz. Foids and quartz cannot form from the same melt.

Nepheline

Chemical Formula: $Na_3K(Al_4Si_4O_{16})$ Crystal System: Hexagonal

Class: Tectosilicate, foid   Crystal Class: Pyramidal

Color: White, gray, light brown, very pale red   Streak: White

Relative opacity: Transparent to opaque

Luster: Vitreous to greasy   Mohs Hardness: 5.5-6, average = 6

Specific gravity: 2.55-2.65, average = 2.59

Fracture: Subconchoidal to uneven   Habit: Massive, granular, crystalline

Formula mass: 146.08 g/mol

Common associated minerals: Alkali feldspar, sodium rich plagioclase, biotite, and pyroxene

Other Characteristics:

Non fluorescent, non magnetic, non radioactive.

It is often confused with transparent to milky quartz. In order to distinguish the two minerals, drop the sample in 5-10% HCl. The nepheline will become cloudy.

Common in alkali rich, undersaturated rocks. It is common in syenite rocks and on geologic maps it may actually be called "nepheline syenite".

Nepheline crystals. A.E. Seaman Mineral Museum.

Sodalite

Chemical Formula: $Na_8(Al_6Si_6O_{24})$

Class: Tectosilicate, foid (sodalite group)

Color: White, gray, light brown, very pale red

Crystal System: Isometric

Crystal Class: Hexatetrahedral

Streak: White if sodalite, bright blue if lazurite, light blue if nosean, very pale blue if hauyne

Relative opacity: Translucent to opaque

Luster: Dull vitreous to greasy   Mohs Hardness: 5.5-6, average = 5.5

Specific gravity: 2.27-2.33, average = 2.30

Fracture: Conchoidal to uneven   Habit: Massive, rare dodecahedral crystals

Molecular mass: 969.21 g

Common associated minerals: Alkali feldspar, sodium rich plagioclase, nepheline, leucite, natrolite, microcline, sanidine, albite, calcite, fluorite, ankerite.

Other Characteristics:

Very fluorescent. Sodalite will fluoresce neon orange in longwave UV light. It will fluoresce yellow in shortwave UV light, non magnetic, non radioactive.

Common in alkali rich, undersaturated rocks. It is commonly forms with other foids. It is its own mineral and also belongs to the Sodalite Group. It is practically impossible to differentiate sodalites without detailed petrographic analysis. Lazurite is a common variety of sodalite and a member of the sodalite group. The Sodalite Group contains; sodalite, hauyne, nosean, and lazurite.

Blue sodalite under visible light (left), and blue sodalite under longwave UV (right). Author's personal collection.

Yooperlite. The neon orange is sodalite fluorescing under longwave UV light. Author's personal collection. Standard photos don't bring out the true colors.

Leucite

Chemical Formula: KAlSi$_2$O$_6$ Crystal System: Tetragonal

Class: Tectosilicate, foid   Crystal Class: Dipyramidal

Color: White, gray, pale yellow to brown hues   Streak: White

Relative opacity: Transparent to translucent

Luster: Vitreous   Mohs Hardness: 5.5-6, average = 5.5

Specific gravity: 2.45-2.50, average = 2.47

Fracture: Conchoidal, uneven

Habit: Euhedral (common), pseudo-cubic as trapezohedron (common), granular (rare), massive (rare)

Molecular mass: 218.25 g

Common associated minerals: Alkali feldspar, sodium rich plagioclase, nepheline, leucite, natrolite, microcline, sanidine, albite, calcite, fluorite, ankerite.

Other Characteristics:

Non fluorescent, non magnetic, non radioactive.

Leucite is the most uncommon of the abundant foids. Present in potassium alkali rich, undersaturated rocks. Leucite will alter to pseudo-leucite (orthoclase, nepheline, analcime) if oxidized. Further oxidation will cause it to become kaolinite clay.

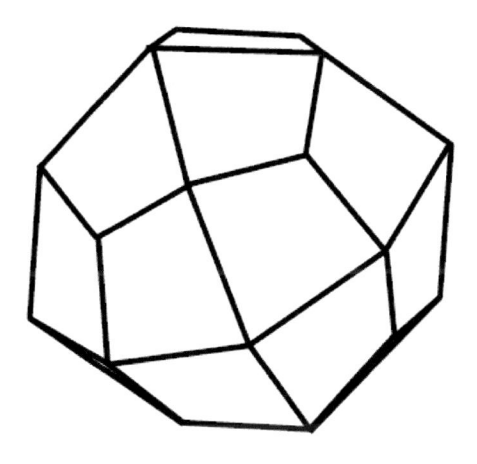

Pseudo-cubic (pseudo-isometric) trapezohedron crystal.

DIORITE VS. GABBRO

Diorite and gabbro are both phaneritic. The two can be hard to differentiate as both are quartz poor and plot in the same area on the QAPF diagram. There are differences.

Diorite

Diorite is of intermediate composition between granites and gabbros. It tends to be rich in felsic minerals as opposed to mafic ones. Felsic plagioclase makes up most of the rock with lesser along with mafics such as amphibole (commonly hornblende), pyroxenes, and minor biotite. By definition diorite is mafic deficient (lacks abundant mafic minerals), compared to gabbro. Diorite tends to have more Na-plagioclase. If the diorite contains significant olivine and iron mafic minerals are present, then the rock is a ferrodiorite, which is transitional to gabbro and very hard to classify. At first glance, diorite tends to have a granitic texture to it.

Common accessory minerals in diorite are zircon, apatite, and magnetite. Diorite rarely occurs as its own plutonic intrusion, but it is commonly in plutons of granite and with gabbro. It can also exist by itself, as dikes and sills. The Nipissing Intrusions of Ontario are commonly diorite and gabbro.

Gabbro

Even though gabbro tends to be mafic rich it can have significant felsic plagioclase, usually in the form of Ca-plagioclase. Gabbro never has a granitic texture. Even though it's phaneritic, it tends to be dark green to dark gray in color.

Gabbros contain the same accessories as diorite with the rare to somewhat common mineral called titanite (a.k.a. sphene). Gabbro rarely forms plutons but occurs commonly as dikes and sills and within other plutonic bodies.

Both diorite and gabbro commonly form in from partial melts of mafic rich rocks above subduction zones.

Below is a photo of a plagioclase rich gabbro I collected and have been slowly polishing over the years. It is common in the Midwest.

This does have quartz in it (yellow arrow pointing to a gray translucent grain of quartz), but it's ~2% of the rock's total volume. This rock is about 43% black hornblende expressed as moderately well developed hexagonal crystal habit (red arrow is pointing to a crystal). Hexagonal is the crystal habit *NOT* the crystal system. Hornblende is monoclinic. This is usually only expressed in well developed crystals. The nearly white mineral is all plagioclase and likely anorthite. The light gray is also plagioclase and likely andesine. Together the two plagioclase make up 55%. There is no measurable alkali feldspar. If we remove the hornblende (it's <50% and is mafic) and normalize our quartz, we get 3.5%. If we normalize our feldspars we get 100% plagioclase. So this is unquestionably a gabbro and it plots about as close as possible to the "P" on the QAP part of the QAPF diagram as you can get.

This picture was taken in Iceland by the author in 2022, during the eruptions at Fagradalsfjall. These eruptions are almost all basalt. Iceland sits on the Mid Atlantic Rift, an active area of plate spreading.

In this photo is one of the many cinder cones created.

VOLUME CALCULATIONS IN PHANERITIC (PLUTONIC) ROCKS

It is very expensive and very time consuming to physically analyze a rock based on its chemical make up. It is also not necessary for most purposes. Most coarse grains (phaneritic) igneous rocks can be classified with relative ease under a hand lens or magnifying glass.

We can't do it by mass, since we are not breaking it down physically. So we do it based on volume of the minerals in the rock in order to get their relative percentages set to 100%. This is a visual procedure and it will generally have a margin of error of about 2-4%. With practice, you will become better at it. Although we say we are getting a volume, we are actually doing it for surface area, since we cannot peak inside the rock.

Important Note: You *ARE* gauging relative mineral percentages based on the visible area they take up in your sample. You are *NOT* counting individual crystals!

I will take you through two examples. When you do this, you are visually inspecting for relative percentages so you can eventually normalize them and plot them on the QAPF diagram.

Sample 1 Analysis

We need to recognize the minerals before we can assign percentages. In this specimen the mafics (dark minerals) are black. Common mafic minerals (and mineral groups) are pyroxene, hornblende, and biotite. Less commonly they are olivine. In this case it's biotite. The alkali feldspars, in this particular case, are pink. The plagioclase (indicated by striations) is white (blue arrow). The quartz is a light gray and translucent (red arrow). Since we have quartz, we know there are no foids, so we will be working strictly with the QAP part of the QAPF diagram. Now that we have identified them, relative percentages can be estimated.

It looks like there are a lot of mafics, but there are only about 17%. Quartz is about 22%. Alkali feldspars are about 35%. This leaves 26% for the plagioclase because all percentages must = 100%.

Now that we have our percentages, we can normalize (p. 67-69) everything to get our rock name. It is a "GRANITE". Specifically a "MOZOGRANITE". Since it contains 15% to <35% biotite as mafics, we can add it to the back of the name. The rock name is: " White mottled pale red with black, MONZOGRANITE with BIOTITE".

Sample 2 Analysis

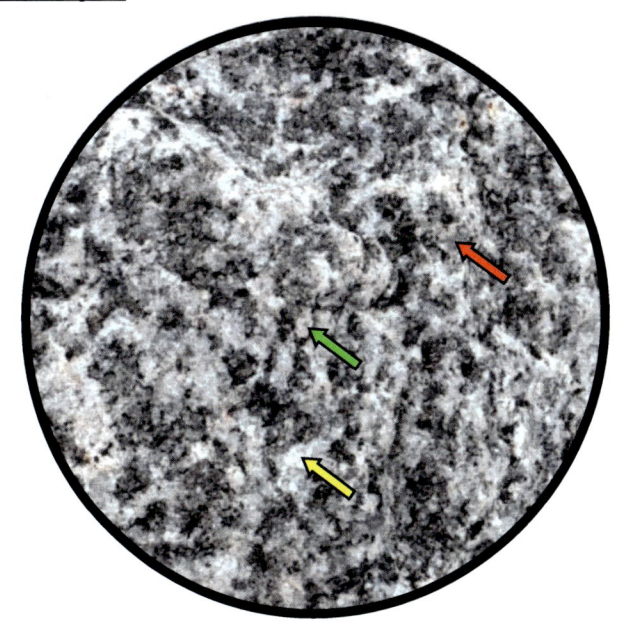

This rock is a little more difficult to name. There is *NO* quartz. There are foids (red arrow) are gray to dark gray. The alkali feldspar and plagioclase are both white, which can complicate things if you are relying strictly on color. You should *NEVER* strictly rely on color. The yellow arrow is the plagioclase as it contains parallel striations. The green arrow is the alkali feldspar.

Mafics are about 39% and are biotite and hornblende. Foids are about 18%. Alkali feldspars are about 23%. This leaves 20% for the plagioclase because all percentages must = 100%.

Now that we have our percentages, we can normalize (p. 67-69) everything to get our rock name. Since we have foids instead of quartz, we will use the APF, or lower part, of the QAPF plot. When we calculate our normalizations just as we would if there were quartz except F stands in place of Q. The rock is a "FOID MONZODIORITE". Since it contains 35-50% biotite and hornblende as mafics (it's hard to tell the exact ratio), we can add it to the front of the name. The rock name is:

"White mottled gray to black, BIOTITE-HORNBLENDE FOID MONZODIORITE".

PART III:
CHARACTERISTICS OF APHANITIC IGNEOUS ROCKS

IDENTIFYING APHANITIC (VOLCANIC) ROCKS

Identifying volcanic (aphanitic) rocks is much more difficult than identifying phaneritic rocks, because crystals are so small generally <0.125mm (the border between very fine and fine sand grains on the Wentworth Scale, see p. 19-20. This is why the volcanic QAPF diagram is much simpler.

With the proper magnifying equipment, crystals closer to the 0.125mm side can be seen. This also means that the rock is likely volcanic. I actually have a minor issue with plutonic and volcanic. I think we should stick with phaneritic and aphanitic, respectively. You can have intrusive basalts. Even some gabbro can be volcanic as in the case with ophiolite (an obducted section of ultramafic ocean crust thrust up onto the continental margin, or in a continental rift zone). So there is a gray area. That is why we should just stick to grain size.

If you can identify the crystals under magnification, then you are good to go, and can use the aphanitic QAPF diagram to get your rock type in the same manner you would for phaneritic rocks. You will also have to normalize (p. 67-69) the percentages, just like for the phaneritic QAPF plot.

What if you can't readily identify the minerals? Are there ways to tell them apart? Yes, to a degree we can use color and density along with some other properties. Ideally you would want to at least look at a thin section or run a chemical analysis. That is not always practical and is not the focus of this book. Relative color is one way, but you need to check it against other identification techniques.

Using Color and a Guide

The following flowchart (next page) is to be used for igneous rocks only. It excludes things like phenocrysts in porphyritic texture or mineral fill in vesicular (amygdaloidal) texture in igneous rock. Certain rocks like andesite and trachyte are nearly always porphyritic to some degree. Rhyolite is not commonly porphyritic, but it can be. There are porphyritic rhyolites near Mountain, Wisconsin. Basalt is almost never porphyritic. Amygdaloidal basalts are very common in the Lake Superior region.

The chart on the next page is derived off the author's experience and education and is first published herein.

Aphanitic (volcanic) Rock Identification Flowchart Based on Color

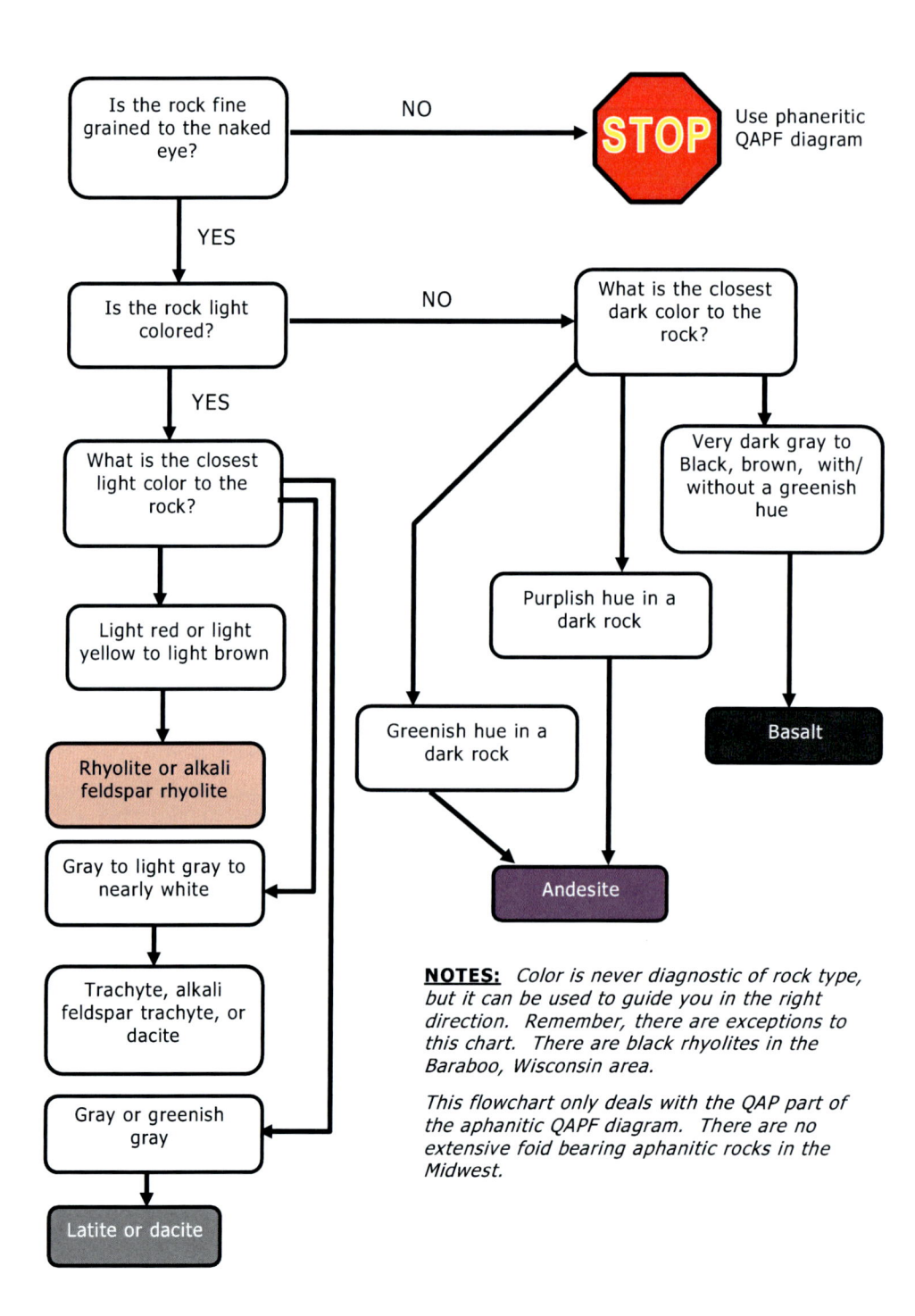

```
[Is the rock fine grained to the naked eye?] --NO--> STOP  Use phaneritic QAPF diagram
         |
        YES
         v
[Is the rock light colored?] --NO--> [What is the closest dark color to the rock?]
         |                                    |                          |
        YES                                   v                          v
         v                        [Purplish hue in a dark rock]  [Very dark gray to Black, brown, with/without a greenish hue]
[What is the closest light color to the rock?]         |                  |
         |                                              v                  v
         v                                    [Greenish hue in a dark rock]   
[Light red or light yellow to light brown]              |              
         |                                              v              
         v                                          Andesite           Basalt
Rhyolite or alkali feldspar rhyolite

[Gray to light gray to nearly white]
         |
         v
[Trachyte, alkali feldspar trachyte, or dacite]

[Gray or greenish gray]
         |
         v
Latite or dacite
```

NOTES: *Color is never diagnostic of rock type, but it can be used to guide you in the right direction. Remember, there are exceptions to this chart. There are black rhyolites in the Baraboo, Wisconsin area.*

This flowchart only deals with the QAP part of the aphanitic QAPF diagram. There are no extensive foid bearing aphanitic rocks in the Midwest.

Porphyritic vs. Amygdaloidal vs. Pegmatitic Textures

It is very important to know the difference between porphyritic, amygdaloidal, and pegmatitic textures. As all three are present in the Midwest.

Pegmatitic Texture

Pegmatitic texture occurs only in plutonic rocks. The rocks are often referred to as pegmatites. Pegmatites are rocks where all (or nearly all) the crystals are >2.54cm (1 inch) in size. This can only occur in a melt that cooled very slowly deep underground. So they do not occur in volcanic rocks, unless they are clasts brought up with a volcanic melt as it came to the surface. This scenario has not been observed in the Midwest. They are most common in granitic rocks, but can exist as dikes or sills in other rocks.

In the photo below, the mostly pink rock is pegmatitic plagioclase (igneous). It exists as two small cross cutting dikes in a gray gneiss (not igneous).

Porphyritic Texture

This is somewhat rare in the Midwest but does occur commonly in andesites and less commonly in rhyolites of the Midwest. Porphyritic rocks are often called porphyry. They are >50% fine grained but will contain isolated to abundant crystals that formed as the melt cooled. This is important, as it distinguishes them from amygdaloidal rocks and xenocryst inclusions. The crystals in porphyritic rocks commonly form first, and are often called phenocrysts. Porphyritic rocks by definition cannot be phaneritic. As it is usually one maybe two of the rock forming crystals that form the phenocrysts. There is no standard as to what size separates the groundmass (fine grains) from a phenocryst. It's just generally accepted that phenocrysts are significantly bigger crystals than the groundmass crystals.

The large porphyry crystals can be any mineral that can solidify from the melt that the rest of the rock did. Commonly, they are made of the same stuff the fine grained majority of the rock is. Quartz, plagioclase, hornblende, and micas. They can aid in identifying the fine grained component of the porphyritic rock. The most common porphyritic rocks in the Midwest are (from most to least common) andesite, rhyolite, dacite, and latite.

Below is a trachyte/phonolite containing (gray groundmass) that contains white to light yellow brown phenocrysts of alkali feldspar and foids. The blue arrows point to some of the phenocrysts. The bright oranges and yellows are lichens on the rock.

Amygdaloidal Texture

This type of texture is very common in the volcanic rocks of the Midcontinent Rift. At first glance, the rock can easily be mistaken for a porphyry. They aren't the same. As an amygdaloidal rocks cool from melt, the lava will often contain gas bubbles. As the gas escapes, leaving cavities called vesicles or vugs (large vesicles). These pockets can later become filled with minerals (kind of like miniature geodes). The minerals may or may not be crystalized from the surrounding groundmass. Hydrothermal fluids and groundwater can introduce foreign chemicals causing minerals to form that are not diagnostic of the groundmass. Most amygdaloidal rocks in the Midwest are basalt. However, andesites can also display this texture.

Amygdaloidal rocks can be distinguished from porphyries easily. There will often be unassociated mineralization in the visible crystals (due to the introduction of foreign chemicals). They are also shaped differently. Phenocrysts will have well defined crystal faces. Amygdaloidal rocks will not. You may get well developed crystal faces within the individual vesicles, but the edges will be rounded because the vesicles themselves are usually spherical, ellipsoid, or deformed versions of ellipsoids in shape. Common minerals filling vesicles and vugs are calcite, quartz, zeolite, epidote, chlorite, or plagioclase. Semi common minerals are prehnite, pumpellyite, thompsonite, and copper minerals.

Below is a cut piece of amygdaloidal basalt (brown purple groundmass) with mineral filled vesicles (dark green and white).

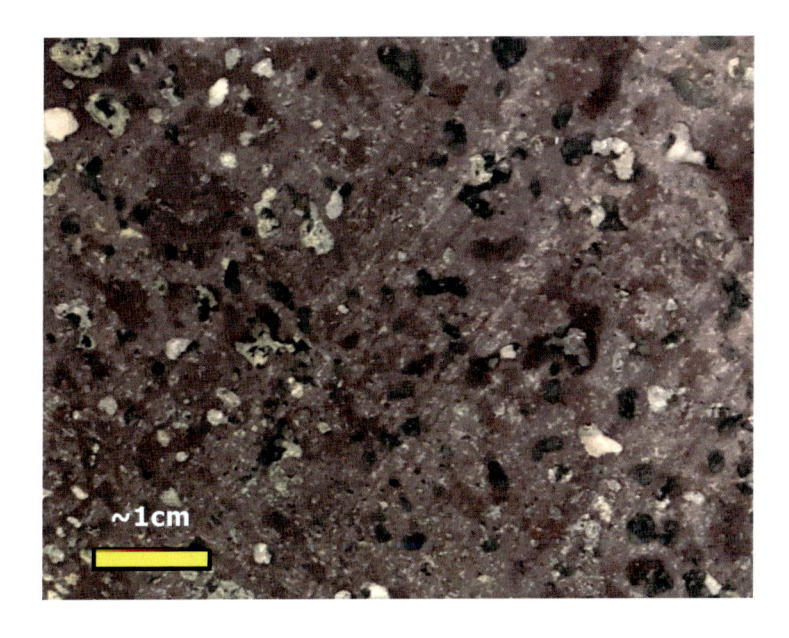

Important Notes on Aphanitic Rock Identification

There are some things you need to keep in mind while identifying aphanitic rocks.

1) Since crystals are often not visible, even under high magnification, color can be used as a guide but is *NOT* diagnostic.

2) Phenocrysts can be useful in identifying the groundmass of a rock, but mineral vesicular fill cannot be used to identify an amygdaloidal rock. Xenocrysts are foreign crystals and small rock fragments brought up in a magma body that that are incorporated into it, but did not melt them before cooling into rock itself.

3) There are very few carbonate igneous rocks in the Midwest. So if you put a drop of weak HCl on a rock and there's a fizzing reaction, you likely don't have an igneous rock. Dark colored fine grained limestones are rare worldwide but common in the Midwest, especially in Pennsylvanian deposits. However, the Mesoproterozoic Nonesuch Formation is a fine grained, often slightly metamorphosed, slate. This can often give it a crystalline appearance. It will often contain some microscopic calcite in the matrix.

4) If you have exhausted all options and can't tell if you have an andesite or a basalt, it's likely basalt. They are more common.

5) The most common volcanic rocks in the Midwest are basalt [to include tholeiite, which the IUGS LeMaitre, et al. (2002) rejects, although *tholeiitic basalt* is acceptable] and andesite. Followed by rhyolite (more common on the edges of the Midcontinent Rift) and dacite. Icelandite appears in the older literature, as in VanSchmus and Hinze (1985). Icelandite plots in between rhyolite, dacite, and andesite/basalt on the QAPF diagram. It can be assigned to any of the aforementioned rock types. There is no practical need for the name in the Midwest.

6) Rhyodacite is the line in between rhyolite and dacite, it serves no real purpose, but you could run across the term.

7) Since basalt and andesite occur in the same part of the QAPF diagram, it is acceptable to name a rock "basaltic andesite" or "andesitic basalt".

8) In rhyolite the quartz is often granular, forming sort of what appears to be similar phenocrysts within the rock, except it is granular without good crystal faces (a.k.a. anhedral).

IUGS QAPF PLOT FOR APHANITIC (VOLCANIC) ROCKS

(2nd order normalization)

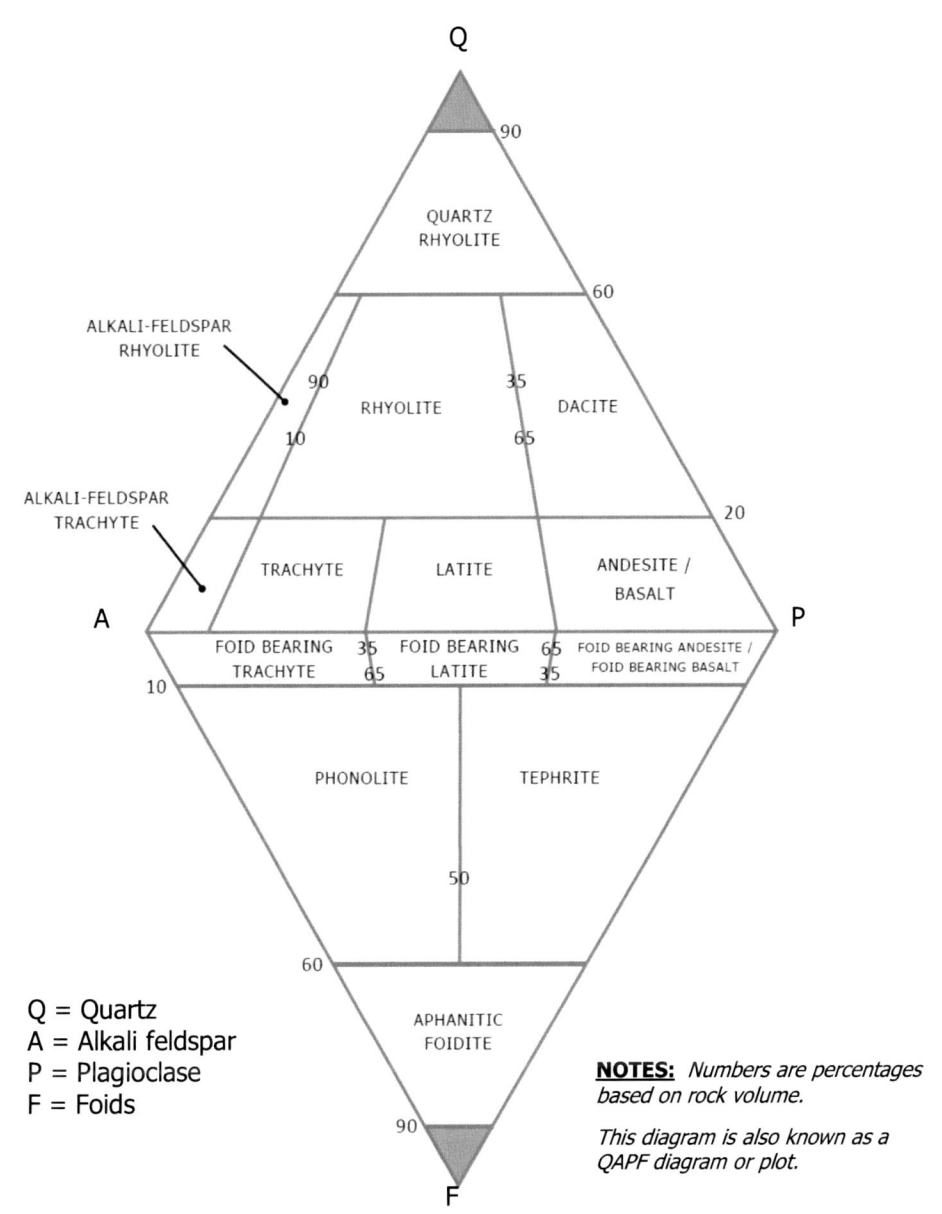

Q = Quartz
A = Alkali feldspar
P = Plagioclase
F = Foids

NOTES: *Numbers are percentages based on rock volume.*

This diagram is also known as a QAPF diagram or plot.

Adapted from: LeBas and Streckeisen (1991), and Gill (2010)

The Aphanitic rock classification ternary diagram is an adaptation from the standards set by the International Union of Geological Sciences (IUGS) **iugs.org**.

This ternary plot is not like the traditional ternary plots used for unconsolidated deposits and sandstones. It uses 3-points but you need to "normalize" (p. 67-69) all three first, and plot your quartz or your foids. Just like you would for any ternary plot.

REMEMBER: *Primary quartz and foids CANNOT crystalize from the same magma. They DO NOT occur in the same rock.*

From here, it differs. Then you plot your quartz or foids as a horizontal line. Once quartz or foids are plotted, then you must re-normalize your plagioclase and alkali feldspars before plotting them. If you normalized them correctly they should be the same line.

Once the two lines are plotted, the point at which they meet is your rock type.

This procedure works exactly the same for phaneritic (i.e. plutonic) rocks (p. 33-34).

NORMALIZING PERCENTAGES IN ROCKS (2nd order normalizations)

Normalization is used in order to utilize the IUGS QAPF plots, for both plutonic and volcanic rocks. Normalizations are also used for metamorphic and sedimentary rocks.

Normalization is (for our purposes), a resetting of the percentage of minerals in a rock based off visual inspection to 100%. Q = quartz, A = alkali feldspar, P = plagioclase. In order to use the QAP diagram, you need to normalize the two percentages to 100%. The same for Q-A-P percentages if mafic and accessory minerals are present.

The equation is exactly the same for each step, you're just using different minerals. We are going to have to do this twice. So we will have a "Part 1" and a "Part 2". You can round as you go (see "notes" on the next page). Try to work with whole number percentages. Decimals aren't really practical using visual methods.

Equation: Part 1)

$$Q = Q_o \left[\frac{100}{(Q_o + A_o + P_o)} \right]$$

Q = Normalized % of quartz
Q_o = Initial % of quartz
A_o = Initial % of alkali feldspar
P_o = Initial % of plagioclase

Once Q is computed, then you can compute A and P separately, using the same equation, except substituting "Q = Q_o" for "A = A_o", then "P = P_o".

Here's an example. Take a felsic plutonic rock that is 5% Mafic, 5% accessory, 15% quartz, 10% alkali feldspar, 65% plagioclase. That will = 100% but for the QAP plot, the mafic and accessory percentages are discarded and we are left with only 90% of the minerals. So now we need to normalize that 90% (or set it to 100%). Plug each mineral into the equation. Your ($Q_o + A_o + P_o$) is going to =90 in this case, because that's the total percent left after the removal of mafic and accessory minerals. I am going to round to whole percentages as I go, using the above equation.

Q = 17%, A = 11%, P = 72%.

Check: 17 + 11 + 72 = 100% so we are now normalized.

We aren't finished. We have to normalize A and P, relative to one another, or it won't plot correctly on the QAP diagram. We calculate using the above equation but leave out Q and Q_o.

Equation: Part 2)

$$A = A_o \left[\frac{100}{(A_o + P_o)} \right]$$

$$P = P_o \left[\frac{100}{(A_o + P_o)} \right]$$

We add $A_o + P_o$. 11 + 72 = 83%, in this case. If we solve for the above 2 equations, A = 13% and P = 87%. 13 + 87 = 100%.

NOW we can plot our quartz, alkali feldspar, and plagioclase amounts! I will use a dotted black line for my A:P ratio and a solid black line for my Q.

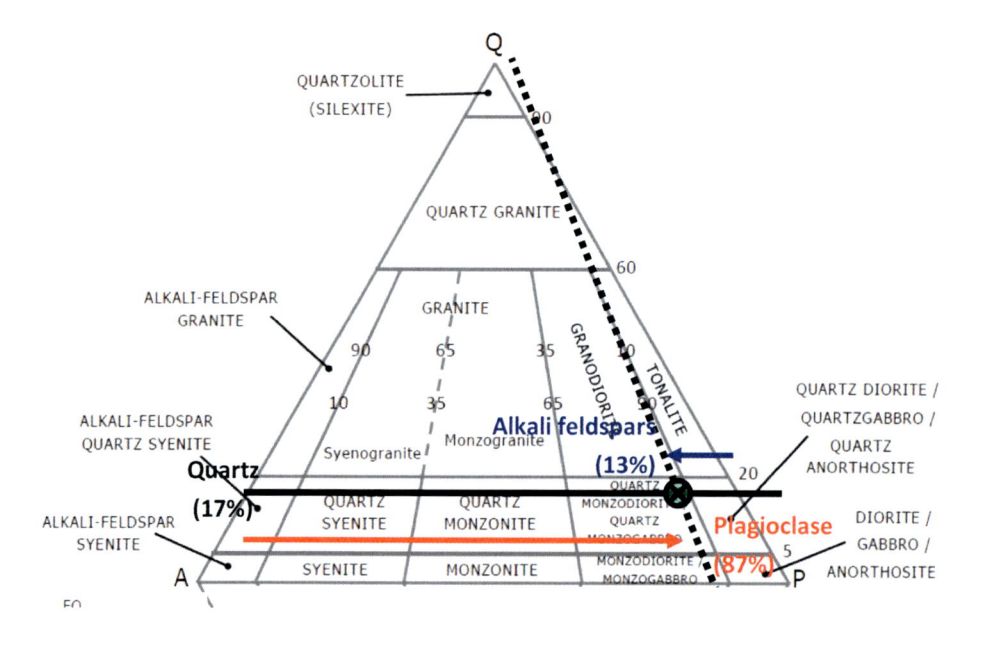

Q = 17%, A = 13%, P = 87%. These 3 do *NOT* add up to 100%, only "A" and "P" do.

You don't need to plot the blue and red arrows. They are just there to show you where to plot the black dotted line. The quartz line is always up from the base in the percentage calculated in your first normalization from "Part 1".

⊗ = Our rock name = monzodiorite. How do we know this? Our example is a felsic plutonic rock. If it had 50- 90% mafics, it would be a monzogabbro.

You would use the same process if foids were present, except you would have F and F_o instead of Q and Q_o. Remember you *CANNOT* have primary quartz and foids in the same rock.

You would also use the same process for volcanic rocks.

NOTES: *While running your calculations for part 1 and your rounding causes your $A_o + Q_o + P_o$ to be 99% or 101%, adjust your QUARTZ percentage up or down to get $A_o + Q_o + P_o$ to = 100%. If this happens for part 2, adjust the percentage plagioclase or alkali feldspar, whichever is more appropriate.*

Normalization Notes:

Normalization is used slightly different than in statistics. Yes, technically these are "quantile normalizations", but with extra steps in the case of second order normalizations.

Herein, 1st order normalizations are when 3-points are plotted on a ternary diagram after removing an unwanted fourth component. In the same way as done on a USDA soil textural chart after removing anything coarser than sand.

Herein, 2nd order normalizations involve two steps, and 2-points are plotted as they are in a QAPF diagram, after removing an unwanted fourth component. Instead of plotting all three points, you normalize for one point and then a second time for the two remaining points that were not calculated in the first normalization.

As for which chart is better, is a matter of opinion. In my opinion, 2nd order normalizations may require an extra step, but they are easier to design and look cleaner.

ARCHEAN VS. PROTEROZOIC IGNEOUS ROCKS (TTG, ANORTHOSITE, AND GRANITE)

This section is just a quick background on something very important as Archean and Proterozoic igneous rocks are very common in Wisconsin, Minnesota, the Upper Peninsula of Michigan, and Ontario.

There are no known Hadean (4.54-4.0Ga) rocks in the Midwest. There are Archean (2.5-4.0Ga) and Proterozoic (0.541-2.5Ga) rocks. Up through most of the Archean, until about 2.7Ga true granites are almost none existent. Instead we have what's referred to as "tonalite-trondhjemite-granodiorite" (TTG) complexes [Hoffman et al. (1988), Frost et al. (2006), and Huang et al. (2013)]. Both tonalite and granodiorite appear on the IUGS QAPF diagram. Trondhjemite is just a very light colored tonalite where almost all of the plagioclase is oligoclase, and it is not necessary to differentiate it from tonalite.

TTG's are *NOT* the same as anorthosite which is almost all plagioclase but the plagioclase is calcium rich, usually anorthite, bytownite, or labradorite. Also remember an anorthosite rock type can but does not have to contain the mineral anorthite. The two are *NOT* synonyms. Anorthosite and TTG's are very common in Ontario and can occur in the same tectonic regime. Anorthosite is rare on Earth (but common on the Moon and possibly Mercury), but they do occur. There are six basic types of anorthosite. Taken from Ashwal, et al. (1983), and Ashwal (2010):

1) Archean mega crystalline anorthosite

2) Proterozoic (massif-type) anorthosite

3) Anorthosite of layered mafic complexes

4) Anorthosite of oceanic settings

5) Other anorthosite

6) Extraterrestrial anorthosite

Although these six types help understand the tectonic regime of anorthosite, they are not part of the descriptive process. Just try and remember that anorthosite can occur with TTG's or granites. But granites do not occur with TTG's, unless tectonics has forced them together. Anorthosite has *NO* volcanic equivalent!

By 2.5Ga TTG's had given way to true granites. They still form today in arc related area and are present in batholiths and in the gabbro sections of ophiolites in subduction zones. Other than that, TTG's are now very rare. Why is this? From the formation of the Earth (~4.54Ga) to about the Paleo-Mesoarchean line (3.2Ga) Plate Tectonics was not in operation. Geologists still argue about what was going on, but it wasn't Plate Tectonics. From about 3.2-2.5Ga, we have a transitioning to Plate Tectonics (Baumann et al. 2015). At 2.5 Ga, we have the Huronian Supergroup (Ontario with equivalents in the Upper Peninsula of Michigan and other places like Wyoming and South Africa, that record a typical Wilson Cycle (the opening and closing of an ocean basin, due to Plate Tectonics).

If you know you are looking at Archean igneous plutonic rocks, there's a really good chance it is either tonalite or granodiorite. Proterozoic rocks would contain granodiorite or granites, but aren't likely to bear TTG's.

GRANOPHYRE

Granophyre, like porphyry, is more of a textural term. Granophyric is actually the accepted textural term. It is not a rock type, which are based off mineral composition. The term is applied to felsic and usually phaneritic rocks with angular euhedral relationships between feldspar and quartz, as well as the mafic minority. It is possible, though rare, for a rhyolite to express granophyric texture. It usually requires a hand lens to see.

The composition of most granophyre rocks is usually alkali-feldspar granite, quartz granite, or syenogranite. Although k-spar is usually dominant, plagioclase granophyre is not unheard of. Rocks as quartz poor and plagioclase rich as diorite can also have granophyric texture.

The texture of a granophyre is expressed as sharp, crystal angles, usually between 45 and 90°. This forms what looks like "V's" all pointing in one direction in sets of offset rows, that will often interlock and change direction.

Granophyre is present in Sudbury Ontario (Lightfoot, 2017).

Typical granophyric texture.

WEATHERING PATTERNS AND IDENTIFYING IGNEOUS ROCKS

Weathered rock can be a great help, or a great hindrance, in identifying rocks. Most geologists don't like identifying weathered outcrops. If the weathering is slight to moderate, it can be a great help to not only identifying structures within the rock, but key in identifying the rock as well. On either a macroscopic or microscopic scale.

We know from Bowen's Reaction Series (p. 26-28) that the first minerals to crystalize are the first to weather out of a rock. They further you go down to the bottom (regardless of continuous or discontinuous series), the more resilient to chemical weathering a mineral is. By the time you get to quartz, only mechanical means can break down the mineral. Quartz reacts essentially with nothing, unless it's re-melted into a new magma.

Since plagioclase can be difficult to discern individual minerals, we use the mafics. Olivine weathers before pyroxene, which weathers before hornblende, which weathers before biotite. How many mafic minerals, and the pattern they produce in the rock, weather we can use this to identify them. This process is hard to explain, so examples would be better.

In this photo of a slightly metamorphosed plutonic/phaneritic rock, the darker minerals can clearly be seen to be weathering out first. For example purposes, we will assume the rock is not metamorphosed at all. The areas of darker mafic minerals are far further recessed (red lines) and the light felsic minerals protrude (green lines).

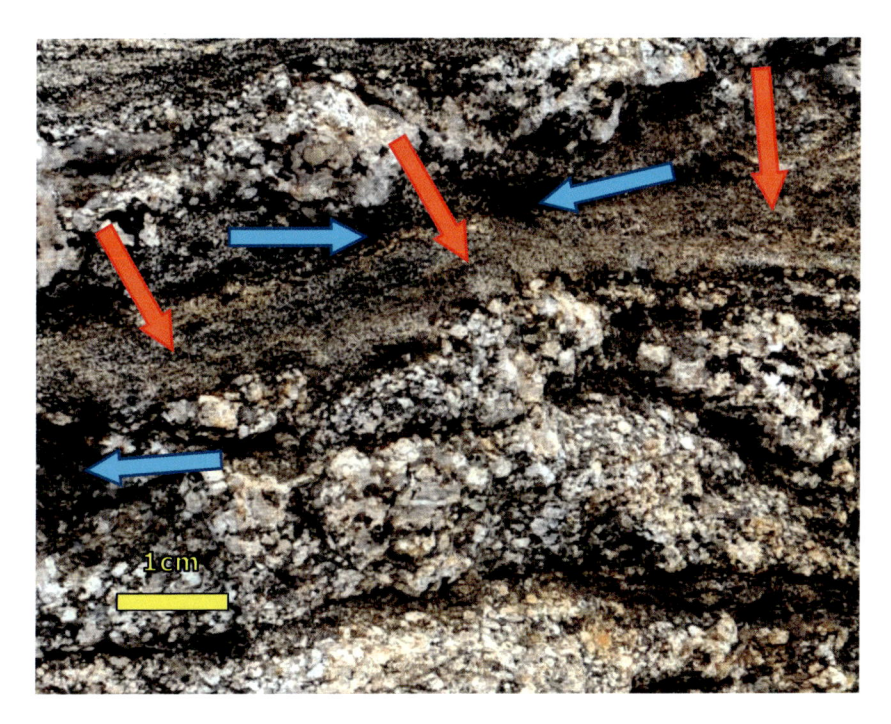

This is a zoom in of the photo on the previous page. Accurate percentages are only possible to about ±5%, as some of the rock has been weathered out. To get better percentages, it will need to be compared to a fresh piece, which will not be done in this example. Let's start with the felsic minerals as those can be identified with ease. This rock is a quartz poor saturated rock, but there is about 10% (light gray translucent grains). The white to pale brown and orange grains are plagioclase, about 60% of the total rock. The slightly yellowish feldspars are alkali feldspar, about 5%. That leaves us with about 25% mafic minerals. Can we identify them? Yes. Since we have abundant plagioclase compared to quartz and feldspar, this tells us the rock cooled within the intermediate to upper felsic part of Bowen's reaction series (p. 26-28). This means our likely mafics are pyroxene and hornblende. The more recessed parts (blue arrows) which would be pyroxene are about 5% of the remaining rock volume. Hornblende (red arrows) makes up the remaining 20% of the rock volume. Now we normalize our percentages (p. 67-69) and calculate our rock type. When we plot that on the QAPF diagram we get a "QUARTZ DIORITE". We have also estimated our two mafics. We do *NOT* normalize our mafic minerals. We can leave off the pyroxene as it's <15%. We have 20% (which falls between 15-35%) hornblende, so we can add it to the end of the name. Our final rock name:
"Brown gray to black mottled white QUARTZ DIORITE with HORNBLENDE".

PYROCLASTIC ROCKS

We have an old saying in geology, "pyroclastic deposits are igneous on the way up and sedimentary on the way down". It is sort of ambiguous to just lump pyroclastics in with igneous rocks, but most of us do. The argument being, if they were not transported too far from their origin and have not incorporated more than 35% of the surrounding rock and sediments as they move, they should still be considered igneous. Fritz (1993) deals with the volcanic vs. sedimentary issue in far deeper detail than I do herein.

The first step in naming a pyroclastic rock, is to determine whether or not the grains are "welded". Welded means that there was enough heat in the clastics that when they were deposited, they weren't melted, but fused together. In my personal experience, I have never seen a pyroclastic in the Midwest that wasn't welded. It is, "welded" is the first word in the name. If the rock isn't welded, you just don't mention it.

Lithic, Crystalline, and Vitric

Next we need to determine whether the rock clasts are mostly lithic (rock fragments), crystalline, or vitric (glass). Glass breaks down within a few million years. All the pyroclastics in the Midwest are billions of years old, so there won't be any vitric material. So you're left with crystalline or lithic. Lithic means the rock consists of non-crystalline grains. Crystalline means just that, there are crystal fragments. Since we won't have glass, basically the ratio of lithic to crystalline needs to be determined. If there's >50% lithics you have a lithic rock. More than 50% crystalline, you have a crystalline rock. I have only found lithic pyroclastic rocks in the Midwest. That doesn't mean crystalline isn't possible. This will be the second word in your pyroclastic name, if you are using welded.

Rhyolitic, Trachytic, Andesitic, Basaltic, and Ultramafic

The third word in our rock name will elude to the chemistry of the rock only if the second word was crystalline, if it can be determined. If the rock is felsic and quartz rich, it'll be rhyolitic. If it's felsic but has no quartz, it'll be trachytic. If it is intermediate to mafic, it'll be either andesitic (intermediate) or basaltic (mafic). If more than 90% mafic, it'll be ultramafic.

One of these words will be the third word in your pyroclastic name, if you can determine the minerology. Even if the rock is crystalline but you can't determine the dominant mineralogy, you just don't mention it.

It may not be possible to get a dominant sense of the rock a gradation if the rock is close to 50% one rock type or another; e.g. you can say, "basaltic-andesitic", or "trachytic-rhyolitic", or "basaltic-ultramafic". What is not possible in a single flow would be a "rhyolitic-ultramafic" pyroclastic rock. You're not going to have a rhyolitic and ultramafic rock from a single source.

Ash, Tuff, Lapilli, Tephra, and Breccia

Your fourth and final word(s) in your name will be either "ash, tuff, lapilli, tephra, or bomb", with the appropriate modifier. For this you need the ternary pyroclastic plots. Tuff is used for pyroclastic flows as where ash is used for pyroclastic sediments. Lapilli and tephra can be really hard to tell apart from sedimentary rock.

Here are some basic definitions first. Ash is non crystalline sedimentary pyroclastic material. Tuff is non crystalline pyroclastic material. Tephra is poorly sorted clastic solid material of all shapes, and lithologies used for pyroclastic sediments. Lapilli is more of a modifying term. It is also used for pyroclastic sediments. Breccia and bombs are >64mm in size. Bombs are usually rounded as where breccia is not.

In my personal experience, I have not seen any Lapilli or tephra pyroclastic rocks in the Midwest.

The pyroclastic rock and sediment charts are used in the same way as most traditional ternary diagrams. There is no need to normalize the percentages. Once you have plotted your rock, you now have the final part of your rock name. This final part should be in all caps and is *NEVER* excluded from the rock name. It is mandatory.

Pyroclastic Rock Classification Diagram

(2nd order normalization)

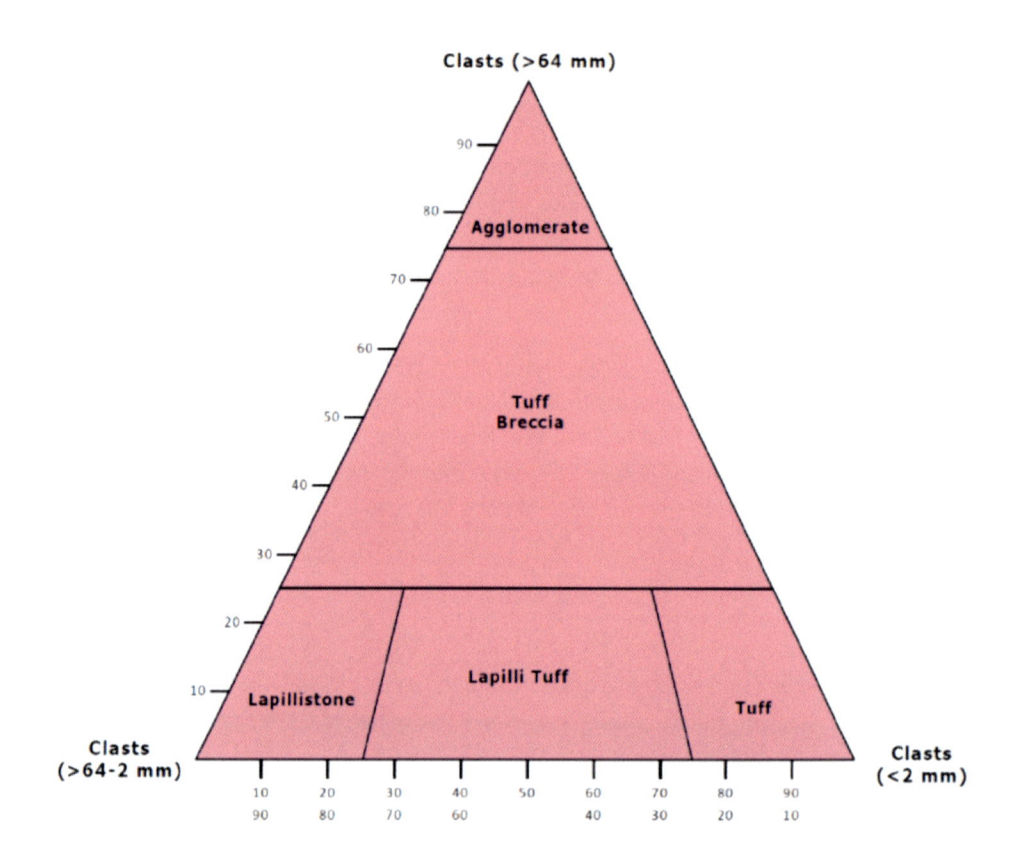

Adapted from: LeBas and Streckeisen (1991), LeMaitre, et al. (2002)

Pyroclastic Sediment Classification Diagram

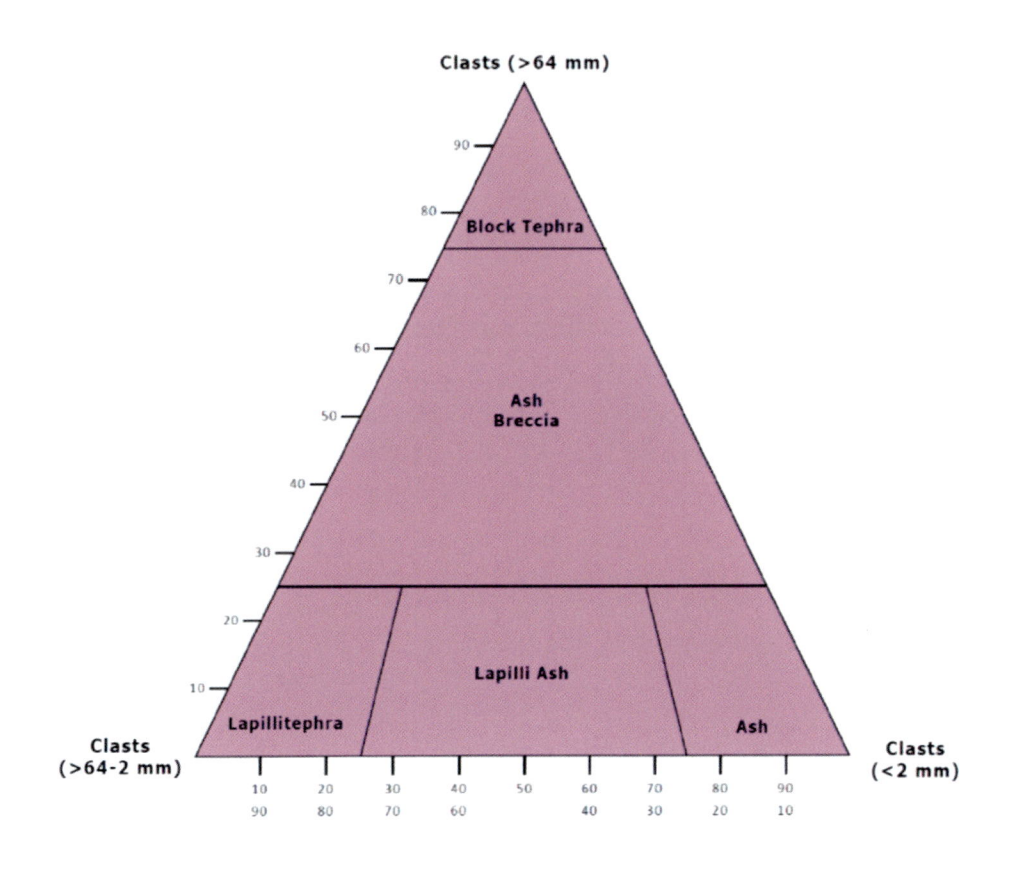

Adapted from: LeBas and Streckeisen (1991), LeMaitre, et al. (2002)

LAMPROPHYRE

Lamprophyres are not common at all but deserve mentioning. They do occur within the Archean rocks as younger intrusions in the Upper Peninsula of Michigan and Ontario. They are not the easiest thing to identify. If you have a nearly black, sort of metallic looking, crystalline rock, on the border of aphanitic and phaneritic, there's a good chance it's a lamprophyre.

They exclusively exist as small dikes and sills in the Midwest and are commonly associated with larger granodiorite intrusions.

They are what is called "ultrapotassic". This just means they have a very large amount of K_2O or NaO_2 in their chemistry. In a practical sense, they are ultramafic.

They are always very dark colors to black and often coarse aphanitic to fine phaneritic in grain size. They will commonly have a metallic sort of luster to them in parts.

Euhedral and essential (primary) minerals are included as phenocrysts. From most to least common phenocrysts you have biotite or phlogopite, hornblende or pargasite, and pyroxene, but NEVER feldspars as phenocrysts.

Minor associated minerals are calcite or zeolite.

Lamprophyres are commonly hydrothermally altered.

See p. 39 for a detailed classification system.

COMPOSITIONAL, TEXTURAL, ROCK NAME BLOCK DIAGRAM

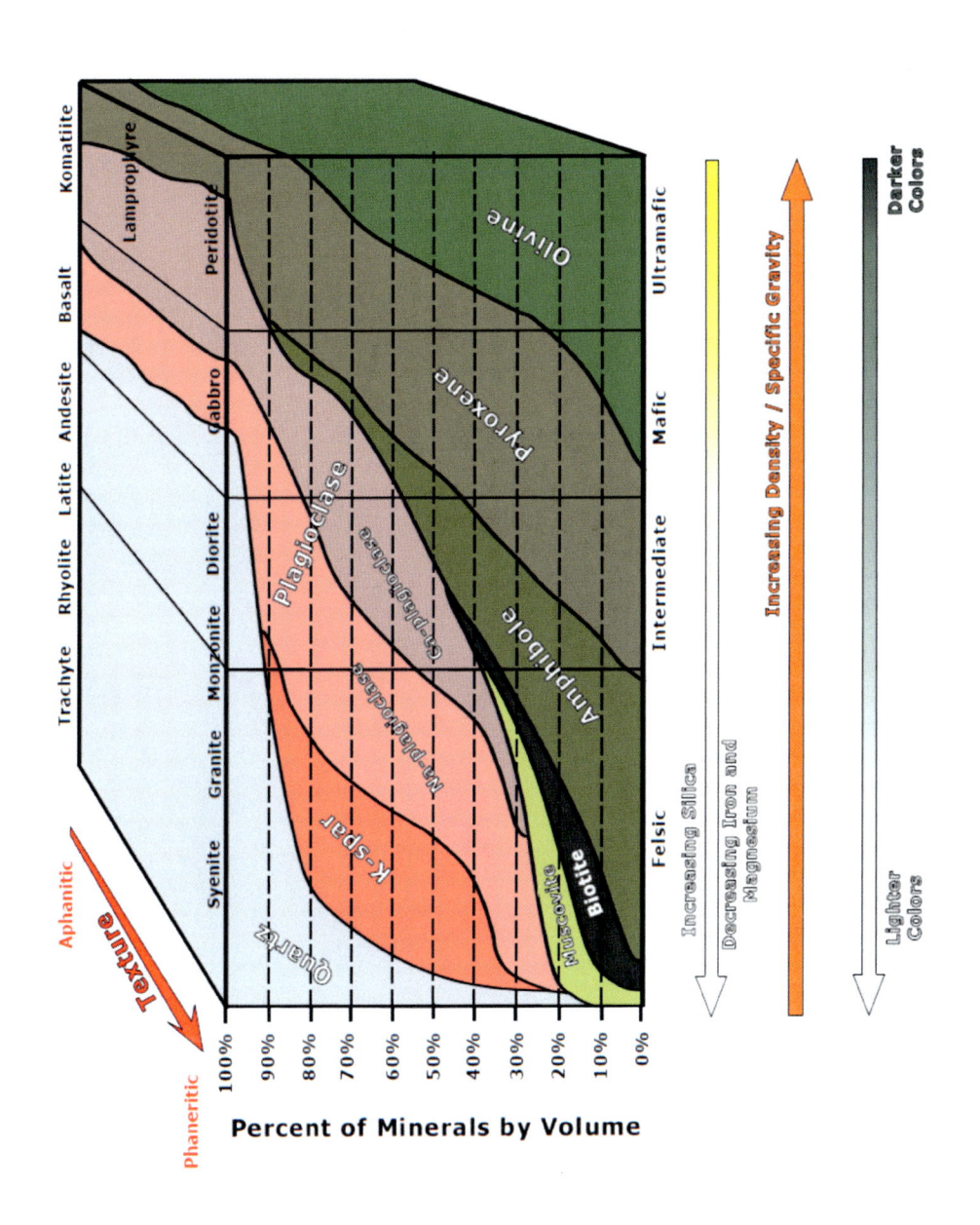

Gill, R., 2010. Igneous rocks and processes, Wiley-Blackwell publishing, ISBN 13:978-1-4443-3065-6
Baumann, S.D.J., 2022. Roadside Geology around the Sudbury area in Ontario, Canada, ISBN 13: 979-8357814425

Relative amounts of minerals in felsic v. intermediate v. mafic v. ultramafic rocks

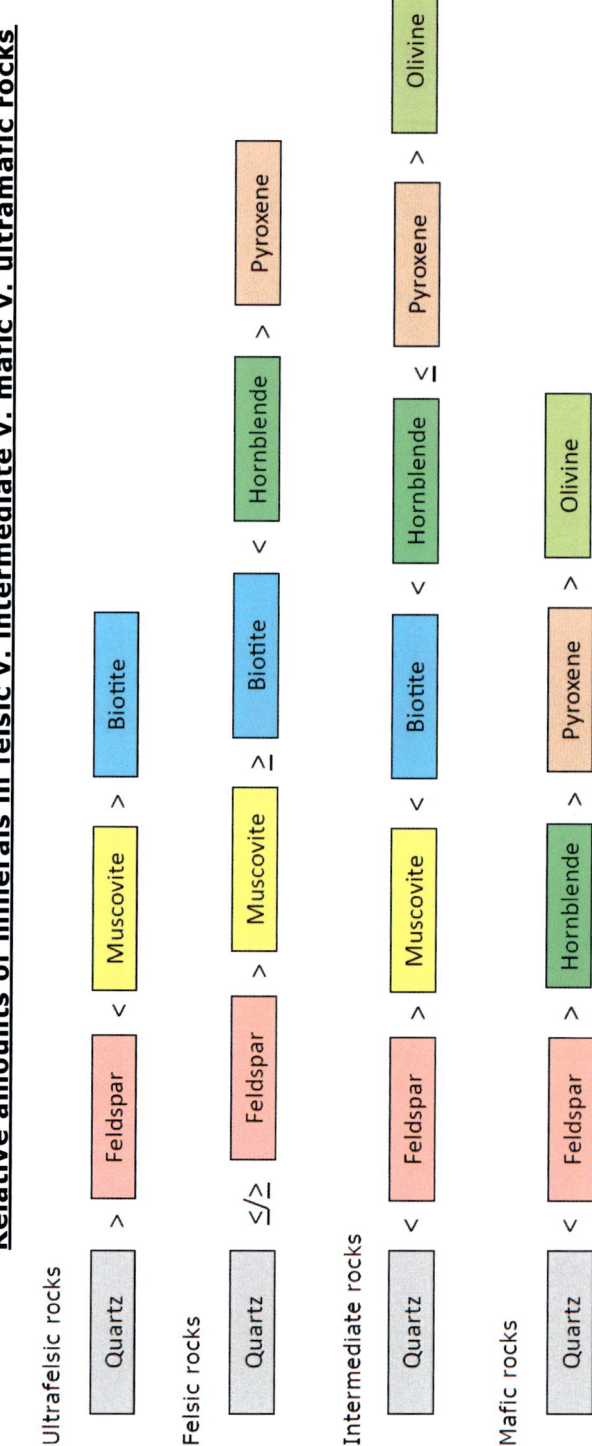

Ultrafelsic rocks

Quartz > Feldspar < Muscovite > Biotite

Felsic rocks

Quartz ≤/≥ Feldspar > Muscovite ≥ Biotite < Hornblende > Pyroxene

Intermediate rocks

Quartz < Feldspar > Muscovite < Biotite < Hornblende ≥ Pyroxene > Olivine

Mafic rocks

Quartz < Feldspar > Hornblende > Pyroxene > Olivine

Ultramafic rocks

Feldspar < Pyroxene ≤/≥ Olivine

Ultrafelsic is named herein. There appear to be ultrafelsig magma bodies in the Michigamme Formation of the Upper Peninsula of Michigan. Any minerals making up <10% of the rock, have been left off this chart. As have accessory or minor rock forming minerals e.g. garnet, epidote, magnetite, etc.

The diagram on p.79 is more for illustration than classification purposes. It shows how the relative concentration of silica (or saturation) in an otherwise homogenous melt, with a set ratio of non silicates, would create certain rocks. It needs to be taken with a grain of salt.

You would use it by inserting a vertical plane through the block and read off the relative percentages.

There are many different things that can effect this diagram, that are not taken into account. For example, if the magma has no potassium in it, you will not get certain minerals (like alkali feldspar) but you would get an anorthosite if other conditions were met. Magma enrichment and depletion is also not taken into account. Nor is incorporation of country rock as the melt cools.

The placements of anorthosite and lamprophyre are somewhat ambiguous. Anorthosite requires a silica poor, relatively mafic poor, magma that consists of almost all the chemicals needed to form plagioclase, even at lower temperatures.

The block diagram does not contain pyroclastic rocks. It focuses on rocks crystalized from melts.

Feel free to use it to help you further understand the crystallization process of magmas and Bowen's Reaction Series (p. 14-17). Use it as a check if you're having trouble identifying a rock by other methods, but never use it to classify the rock.

SIMPLIFIED CLASSIFICATION COMPARISON CHART

Composition	Felsic	Intermediate	Mafic	Ultramafic
Primary mineral assemblage	Quartz, K-spar, Na-plagioclase	Hornblende, Na-Ca plagioclase	Pyroxene, Ca-plagioclase	Olivine, Pyroxene
Accessory mineral assemblage	Biotite, Muscovite	Pyroxene, Biotite	Hornblende, Olivine	Ca-plagioclase
Phaneritic (coarse grained)	**Granites**	**Diorite**	**Gabbro**	**Peridotite**
Aphanitic (fine grained)	**Rhyolite**	**Andesite**	**Basalt**	**Lamprophyre**
Porphyritic (fine grained with phenocrysts)	**Porphyritic Rhyolite**	**Porphyritic Andesite**	**Porphyritic Basalt**	**Komatiite**
Glass (microcrystalline)	**Obsidian and Pumice**			
Pyroclastic (fragment grained)	**Tuff and Breccia**			

Net Rock Color: LIGHT → DARK

Rock names are in black bold type.

Adapted from: *geologyin.com*

The diagram on the previous page is more for illustration than classification purposes. It shows the rock names drawn from comparison with all the primary and accessory mineral assemblages, and in relationship to all the textures of igneous rocks along with general color.

This chart includes pyroclastic rocks not just rocks crystalized from melts.

Feel free to use it to help you further understand the crystallization process of magmas and Bowen's Reaction Series (p. 26-28). Use it as a check if you're having trouble identifying a rock by other methods, but never use it to classify the rock.

PROPERTIES OF COMMON MAFIC MINERALS

Mafic minerals that appear in rocks exist by the dozens. However, there are only a few that are common.

Olivine

Olivine is a group of minerals but olivine itself is a mineral within the group and the most common.

Chemical Formula: $(Mg,Fe)_2SiO_4$ Crystal System: Orthorhombic

Crystal Class: Not applicable

Color: Yellow, green yellow, bright green

Streak: None

Relative opacity: Transparent to translucent

Luster: Vitreous   Mohs Hardness: 6.5-7

Specific gravity: 3.2-4.5

Fracture: Conchoidal to brittle   Habit: Massive, granular

Molecular mass: 153.31g

Common associated minerals: Magnetite, hornblende, augite

Other Characteristics:

Mg rich olivine is not stable in the presence of silica. It reacts with it to form orthopyroxene $(Mg, Fe)_2Si_2O_6$.

Olivine (green mineral) with a U.S. penny for scale.

Hornblende

Hornblende isn't a mineral in and of itself, but is a closely related series of minerals with varying amounts of Fe and Mg. Na can substitute for Ca. Hornblende is also generically used to refer to all amphiboles, and it is a part of the amphibole group. However, out of the dozen plus minerals in the amphibole group, hornblende is the only common one in igneous rocks. Minerals in the amphibole group are either orthorhombic or monoclinic.

Chemical Formula: $Ca_2(Mg,Fe, Al)_5(Al,Si)_8O_{22}(OH)_2$ Crystal System: Monoclinic

Crystal Class: Not applicable

Color: Dark green to black

Streak: Light gray to nearly white (ferrohornblende), white to colorless (magnesiohornblende)

Relative opacity: Opaque

Luster: Vitreous to dull   Mohs Hardness: 5-6

Specific gravity: 2.9-3.4

Fracture: Uneven to splintery   Habit: Granular, pseudo-hexagonal

Molecular mass: 821.16g

Common associated minerals: Hedenbergite (in granites), albite, biotite, epidote, quartz

Other Characteristics:

The iron rich end of hornblende is ferrohornblende and the magnesium rich end is magnesiohornblende.

Augite

Augite is the most common of the pyroxene (p. 87-93) minerals in igneous rocks. It belongs to the most abundant clinopyroxene minerals, which are all monoclinic (p. 88). Other common clinopyroxene minerals are pigeonite (common in lamprophyres), and diopside (common in kimberlites). Orthopyroxene minerals are orthorhombic (p. 87) and the most common one in igneous rocks (and to a lesser degree in metamorphic rocks) is hypersthene.

Chemical Formula: $(Ca,Na)(Mg,Fe,Al,Ti)(Si,Al)_2O_6$ Crystal System: Monoclinic

Crystal Class: Prismatic

Color: Brown (most common), brown green, brown purple, black

Streak: Very pale greenish to nearly white

Relative opacity: Transparent to opaque

Luster: Vitreous to dull   Mohs Hardness: 5.5-6

Specific gravity: 3.19-3.56 (3.38 average)

Fracture: Uneven to conchoidal   Habit: Stubby prisms, skeletal, dendritic

Molecular mass: 236.35g

Common associated minerals: Orthoclase, labradorite, olivine, leucite, sanidine, amphiboles, and other pyroxenes

Other Characteristics:

An essential mineral in gabbro and basalt.

Pyroxene

Pyroxene is not a mineral. It is a group of similar minerals that is often generically lumped together in the same way amphiboles are. Pyroxene is abbreviated Px. There are two crystal systems that pyroxenes can be in. Clinopyroxene (Cpx or CPx), which are monoclinic. You have orthorhombic pyroxenes, which are orthopyroxene (Opx or OPx). Pyroxenes have a general chemical formula is $XY(Si,Al)_2O_6$. Where X = Ca, Na, Mg, Fe (II) or rarely Li, Mn, Zn. Y is a bit more complicated; but generally Y = Co, Cr, Fe (III), Mg, or Mn.

Since there are nearly two dozen pyroxene minerals, I will not be including lengthy individual mineral data sheets. Below is a chart that depicts some easy tests that can be used to differentiate common pyroxenes.

Theoretically wollastonite would be a calcium end member of Ca-Mg-Fe pyroxenes, it is considered a metamorphic mineral (derived from carbonates) that is not technically a pyroxene, and thus it is not included. Ferrosilite is an Opx an is not included, as it to is metamorphic. Ferrosilite forms in medium to high grade metamorphosed iron formations. Hedenbergite is a common clinopyroxene but only in metamorphic skarn and metamorphosed iron formations, so it is not included. Jadeite is a common clinopyroxene but only forms under high metamorphic temperature conditions with under low pressure, so it is not included. SG = specific gravity.

Orthopyroxene (Opx): Orthorhombic

Mineral	Chemical Formula	Mohs Hardness	Streak	Other
Enstatite	$MgSiO_3$	5-6	Gray	SG = 3.2-3-3
Hypersthene	$(Mg,Fe)SiO_3$	5-6	Very pale gray to pale green gray	SG = 3.4-3-9 Vitreous to pearly luster

Clinopyroxene (Cpx): Monoclinic

Mineral	Chemical Formula	Mohs Hardness	Streak	Other
Augite	----------	------	------	See p. 22,86,122
Aegirine	$NaFe^{3+}Si_2O_6$	6	Yellow gray	SG = 3.50-3.60 Commonly forms with K-spar and foids.
Clinoenstatite	$Mg_2Si_2O_6$	5-6	White	Forms commonly as glassy phenocrysts in high magnesium to nearly mafic andesites.
Clinoferrosilite	$(Fe^{2+},Mg)_2Si_2O_6$	5-6	White	Glassy appearance
Diopside	$MgCaSi_2O_6$	5.5-6.5	White	SG = 3.278 Ultramafic in peridotite and kimberlite rocks.
Omphacite	$(Ca;Na)(Mg; Fe;Al)Si_2O_6$	5-6	Pale green to nearly white	SG = 3.16-3.43 Forms anhedral to slightly prismatic crystals.
Pigeonite	$(Mg;Fe^{2+};Ca)(Mg;Fe^{2+})Si_2O_6$	6	Pale gray to nearly white	SG = 3.17-3.46 Common in silica rich igneous rocks. Occurs with augite and olivine.
Spodumene	$LiAlSi_2O_6$	6.5-7	White	SG = 3.03-3.23 Common in aplite and lithium rich pegmatitic granite. Occurs with albite, lepidolite (bright pink mica), and beryl.

Accepted pyroxene chemical subdivisions

By definition, pyroxenes (this excludes the pryroxenoids) must fit the following criteria:
1) They fall into either monoclinic or orthorhombic crystal systems
2) Their structures are single *NOT* double chains of silicate tetrahedra
3) They are anhydrous meaning they do *NOT* contain any OH or H_2O in their chemical formulas

Mg-Fe Pyroxenes				
Mineral Name	OPx, CPx, Pxoid	Crystal System	Chemical formula	End Member Composition
Ferrosilite	OPx	Orthorhombic	$(Mg,Fe)_2Si2O_6$	$Fe^{2+}_2Si_2O_6$
[2]Eulite	OPx	Orthorhombic	$FeSiO_3$	N/A
Hypersthene	OPx	Orthorhombic	$(Mg,Fe)SiO_3$	N/A
[1]Bronzite	OPx	Orthorhombic	$(Mg,Fe)_2Si2O_6$	N/A
Enstatite	OPx	Orthorhombic	$MgSiO_3$	$Mg_2Si_2O_6$
[3]Clinoenstatite	CPx	Monoclinic	Mg_2SiO_6	N/A
[4]Clinoferrosilite	CPx	Monoclinic	$Fe^{2+}_2Si_2O_6$	N/A
Pigeonite	CPx	Monoclinic	$(Ca,Mg,Fe^{2+})(Mg,Fe^{2+})Si_2O_6$	N/A

[1]Bronzite is the intermediate between enstatite and hypersthene
[2]Eulite is the intermediate between hypersthene and ferrosilite, it is the simplified chemical form of ferrosilite
[3]Clinoenstatite is a polytype of enstatite
[4]Clinoferrosilite is a polytype of ferrosilite

Mn-Mg Pyroxenes

Mineral Name	OPx, CPx, Pxoid	Crystal System	Chemical formula	End Member Composition
Donpeacorite	OPx	Orthorhombic	$(Mn,Mg)MgSi_2O_6$	N/A
Kanoite	CPx	Monoclinic	$(Mn,Mg^{2+})Si_2O_6$	$MnMgSi_2O_6$
*Nchwaningite	Pxoid	Orthorhombic	$Mn^{2+}SiO_3(OH)2 \cdot (H_2O)$	N/A
Pyroxmagite	Pxoid	Triclinic	$Mn^{2+}SiO_3$	N/A
Rhodonite	Pxoid	Triclinic	$(Mn^{2+}Fe^{2+}Mg,Ca)SiO_3$	N/A

* Nchwaningite (Nyfeler and Dixon, 1995) named the mineral and referred to it as "pyroxene like". By definition, pyroxenes do not contain OH or H_2O. Herein it is included as a pyroxenoid.

Ca Pyroxenes

Mineral Name	OPx, CPx, Pxoid	Crystal System	Chemical formula	End Member Composition
Augite	CPx	Monoclinic	$(Ca,Na)(Mg,Fe,Al,Ti)(Si,Al)_2O_6$	N/A
Diopside	CPx	Monoclinic	$Ca(Mg,Fe)Si_2O_6$	$Ca,MgSi_2O_6$
Esseneite	CPx	Monoclinic	$Ca,Fe^{3+}Al,SiO_6$	$Ca,Fe^{3+}Al,Si_2O_6$
Hedenbergite	CPx	Monoclinic	$Ca,Fe^{2+}Si_2O_6$	$Ca,Fe^{2+}Si_2O_6$
Johannsenite	CPx	Monoclinic	$(Ca,Mn)Si_2O_6$	Ca,Mn,Si_2O_6
Petedunnite	CPx	Monoclinic	$Ca(Zn,Mn^{2+},Fe^{2+},Mg)Si_2O_6$	Ca,Zn,Si_2O_6
Wollastonite	Pxoid	Triclinic / *Monoclinic	$CaSiO_3$	$CaSiO_3$

* polytype

Pale yellow highlighted minerals are the most common

Adapted from: *Morimoto et al., 1989*

Ca-Na Pyroxenes

Mineral Name	OPx, CPx, Pxoid	Crystal System	Chemical formula	End Member Composition
Aegirine-augite	CPx	Monoclinic	$(Ca,Na)(Mg,Fe^{2+},Fe^{3+})Si_2O_6$	N/A
Omphacite	CPx	Monoclinic	$(Ca,Na)(Mg,Fe,Al)Si_2O_6$	N/A

Na Pyroxenes

Mineral Name	OPx, CPx, Pxoid	Crystal System	Chemical formula	End Member Composition
Aegirine	CPx	Monoclinic	$Na,Fe^{3+}Si_2O_6$	$Na(Al,Fe^{3+})Si_2O_6$
Jadeite	CPx	Monoclinic	Na,Al,Si_2O_6	$Na(Al,Fe^{3+})Si_2O_6$
Jervisitite	CPx	Monoclinic	$Na,Sc^{3+}Si_2O_6$	N/A
Kosmochlor	CPx	Monoclinic	$Na,Cr^{3+}Si_2O_6$	N/A

Na-Mn Pyroxenes

Mineral Name	OPx, CPx, Pxoid	Crystal System	Chemical formula	End Member Composition
Namansilite	CPx	Monoclinic	$Na,Mn^{3+}Si_2O_6$	N/A

Li Pyroxenes

Mineral Name	OPx, CPx, Pxoid	Crystal System	Chemical formula	End Member Composition
Spodumene	CPx	Monoclinic	Li,Al,Si_2O_6	N/A

Adapted from: *Morimoto et al., 1989*

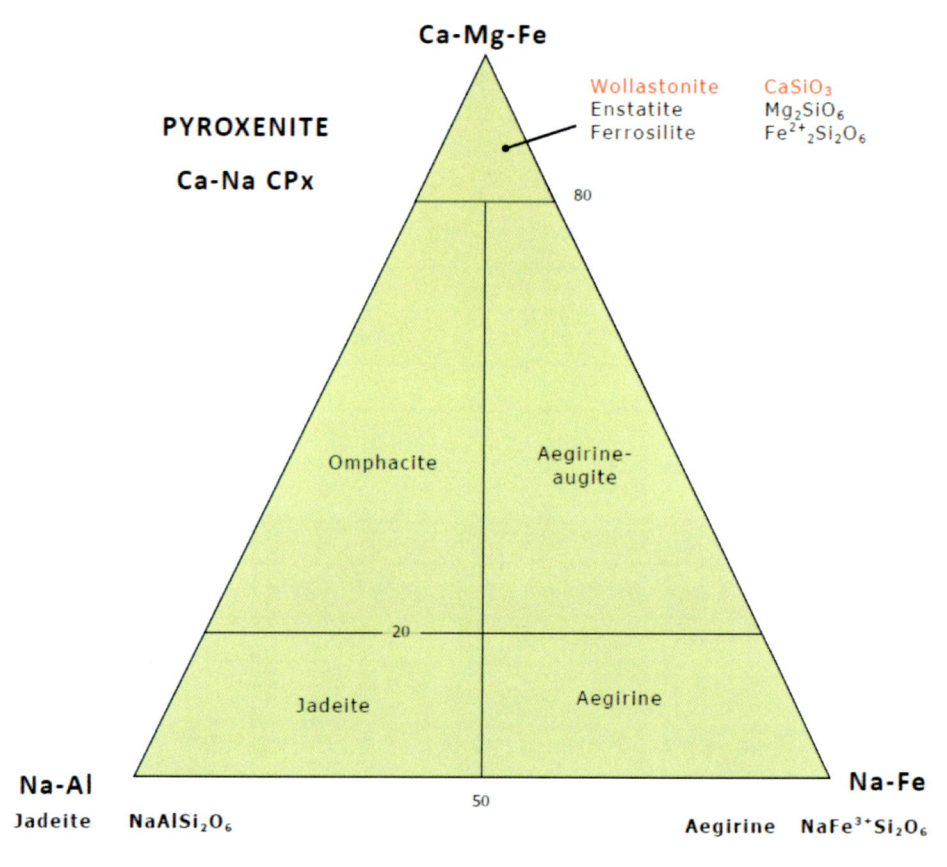

Ca-Mg-Fe

PYROXENITE

Ca-Na CPx

	Wollastonite	$CaSiO_3$
	Enstatite	Mg_2SiO_6
	Ferrosilite	$Fe^{2+}{}_2Si_2O_6$

- 80
- Omphacite
- Aegirine-augite
- 20
- Jadeite
- Aegirine
- 50

Na-Al — Jadeite $NaAlSi_2O_6$

Na-Fe — Aegirine $NaFe^{3+}Si_2O_6$

Wollastonite ($CaSiO_3$) is a pyroxenoid not pyroxene. Wollastonite is structurally the same and contains nearly 100% Ca as its non silicate chemical components. The only significant difference is the Ca abundance. Many Px have Ca in them, just not as much as Wollastonite. Plus there's one other property that separates it from the true Px. Wollastonite can form monoclinic or triclinic crystals. As where all pyroxenes form monoclinic (CPx) or orthorhombic (OPx) crystals.

(2nd order normalization)

Adapted from: Morimoto et al, 1989

PYROXENITE
Ca-Mg-Fe CPx

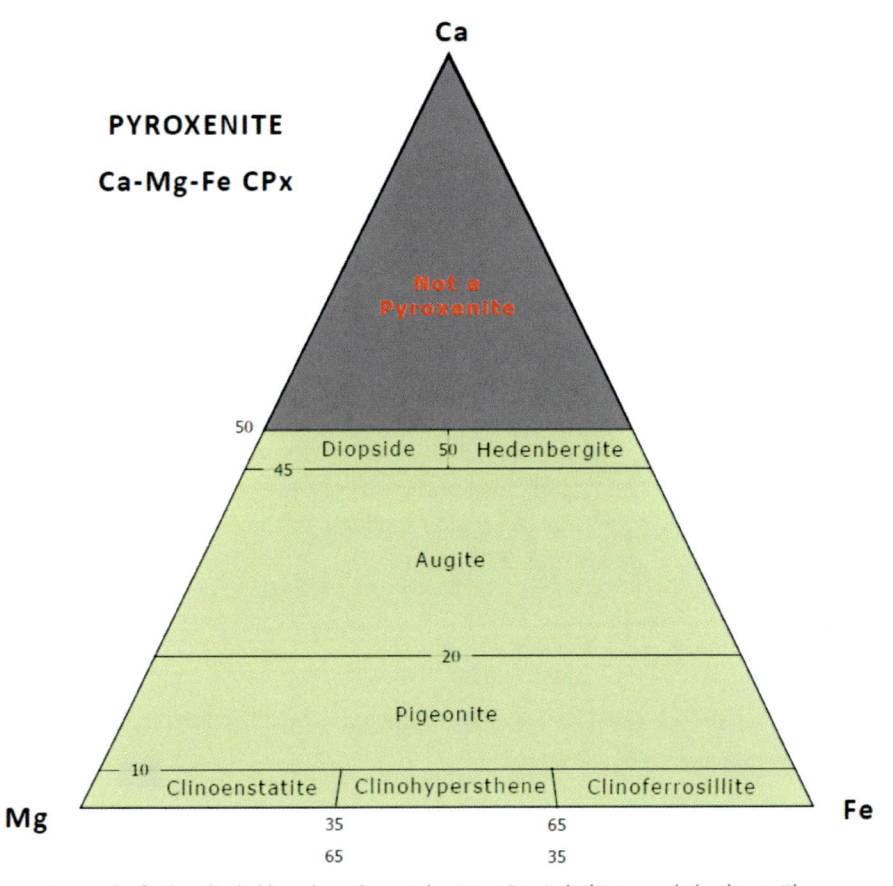

Ca

Not a Pyroxenite

50 — Diopside — 50 — Hedenbergite
— 45 —

Augite

— 20 —

Pigeonite

— 10 —
Clinoenstatite | Clinohypersthene | Clinoferrosillite

Mg — 35 — 65 — Fe
65 — 35

Pyroxenites can be further divided based on elemental ratios. Granted, this can only be done with chemical analysis. If you have that ability, feel free to use this chart. Anything plotting >50% Ca is a carbonatite, not a pyroxenite.

This chart is a vatiation of the "Subcommitee on Pyroxenes" chart. Some of you may be familiar with Polervaart and Hess' 1951 ternary plot. But many of the names have been abandoned. Such as Ferrosilite, Ferroaugite, ferrohedenbergite, salite, endiopside, magnesium pigeonite, etc. I did keep the name "clinohypersthene" because it is the only intermediate from Polervaart and Hess that didn't overcomplicate things.

There is a CPx mineral that does not fit on this graph because it contain large amounts of manganese (Mn). It falls under the "Mn-Mg" CPx minerals (Morimoto et al, 1989). The mineral is:

Namansilite $NaMn^{3+}Si_2O_6$

(2nd order normalization)

Adapted from: Morimoto et al, 1989

NATIVE ELEMENTS AND SIMILAR LOOKING MINERALS

There are many native elements in the Midwest. Perhaps the most abundant is native copper. Zinc, silver, and gold also occur but to a much lesser degree. First I am going to deal with copper, since it is the most common. Then I will get into fool's gold (pyrite and chalcopyrite) and how to tell the two apart, along with gold (p. 96-97).

Copper

Native copper isn't usually mistaken for other minerals due to its unique metallic orange to orange red color. It also oxidized a bright blue green to light green. So it's pretty easy to pick out. It is very common in the Mesoproterozoic Midcontinent Rift volcanics and interbedded sediments of the Portage Lake Volcanics and its equivalents. It also occurs in both native and disseminated copper in the overlying Mesoproterozoic sediments of the Oronto Group. It also commonly occurs in the Archean TTG rocks.

Chemical Formula: Cu   Crystal System: Isometric

Color: Metallic orange to orange red, oxidized bright blue green to light green

Streak: Metallic orange to orange red

Relative opacity: Opaque

Luster: Metallic   Mohs Hardness: 2.5-3

Specific gravity: 8.93-8.95 Density (solid): 8.94g/cm^3 Density (liquid): 8.02g/cm^3

Fracture: Hackly   Habit: Massive, cubic, dodecahedra, tetrahexahedra, skeletal, dendritic

Molecular mass: 63.55g

Melting point: 1084.62°C (1984.32°F)   Boiling Point: 2562°C (4643°F)

Common associated minerals: Silver, chalcocite, chalcopyrite, cuprite, azurite, malachite, iron oxides

Other Characteristics:

Malleable and commonly exists in pores of volcanic rocks and in between sediment grains.

Native copper on white quartz. U.S. penny for scale. Upper Peninsula of Michigan.

Borderline native-disseminated copper in white quartz which serves as the matrix for a mafic breccia. U.S. penny for scale. Upper Peninsula of Michigan.

Gold vs. Pyrite and Chalcopyrite

In the Midwest, gold commonly occurs in the Archean greenstones of Ontario. Unfortunately for those seeking it, so do sulfides (pyrite, chalcopyrite, etc.) that closely resemble it. Sometimes called "fool's gold". Instead of giving a spread of each mineral, I am going to do a comparative chart (see below). Of the three minerals below, chalcopyrite is by far the most common. Chalcopyrite commonly occurs with copper as copper is in its chemical formula. Chalcopyrite is a major ore of copper.

Characteristic	Gold	Pyrite	Chalcopyrite
Color	Silvery yellow to golden	Pale brass yellow to golden	Brass yellow to golden
Chemical Formula	Au	FeS_2	$CuFeS_2$
Tarnish	No	Yes (darker)	Yes (metallic red-purple hues or
Crystal system	Isometric	isometric	Tetragonal
Crystal habit	Nuggets, flakes, small grains, sheets, and very rarely cubic or octahedral crystals	Massive, radiated, granular, stalactitic, cubic, octahedral, pyritohedrons	Massive, botryoidal, tetrahedral
Striations present	No	Yes, but not always present	No
Mohs hardness	White: 2.8-4 14k yellow: 3-4 18k yellow: 2.75	6-6.5	3.5
Density	$19.32 gm/cm^3$	$5.01 g/cm^3$	$4.19 g/cm^3$
Odor when	No	Yes, sulfur or metallic	Not usually
Weathering pattern	Mechanically breaks down, it essentially doesn't oxidize and is non-reactive	Chemically breaks down rusty or yellowish	Chemically breaks down into other copper minerals, usually bright blue or blue
Luster	Metallic	Metallic	Metallic

Characteristic	Gold	Pyrite	Chalcopyrite
Magnetic	No (unless there are impurities such as iron, nickel, cobalt, gadolinium, dysprosium)	Paramagnetic (weakly attracted to a pole magnet but not permanently magnetic)	Magnetic *ONLY* when heated
Malleable	Yes	No	No
Fracture	None	Uneven sometimes conchoidal	Irregular to uneven
Streak	Golden yellow	Black green to brown green	Black green
Opacity	Opaque	Opaque	Opaque
Mineral class	Native element	Sulfide	Sulfide

Some of the information in this chart was retrieved from: The **Utah Department of Natural Resources,** the **Utah Geological Survey** division at, geology.utah.gov/popular/general-geology/rocks-and-minerals/utah-gold/fools-gold/ and geology.utah.gov/popular/general-geology/rocks-and-minerals/utah-gold/fools-gold/

AN EXAMPLE OF CLASSIFYING IGNEOUS ROCKS

I have decided not to make a specific and ridged standard. There are some things that should always be included at minimum. They are in the list is below. Plus I will walk you through one complete example. This list can be expanded upon based on the detail of the petrographic analysis, time constraints, or desired detail.

Minimum Requirements for Naming an Igneous Rock

1) Color. Of the rock as a whole. You can break down colors further if you wish.

2) Significant modifier. Either before the body of the rock name (if 35% to 50% of the rock). Or after followed by "with" (15% to <35%) of the rock. At the very end of the name the words "minor" can be added if the rock contains 5% to <15% minor minerals. The most optional is "trace". Trace would be included after "minor" and is used only for distinctive rock defining mineral <5% of the rock. Trace should be excluded in the rock name if it doesn't aid in differentiating the rock from other rocks.

3) These percentage standards can vary slightly. If a rock is only 12% hornblende, but the crystals are large, you can add them to the name. If they are fine grained, you probably shouldn't add them to the name. If a rock were 53% coarse aphanitic to fine phaneritic grained pyroxene, within coarse or pegmatitic plagioclase crystals, you may want to add it to the name because of the stark contrast, and use the plutonic QAPF diagram, not the mafic classification even though mafics are >50%.

4) Body of the rock name. Derived from either the phaneritic QAPF diagram, the aphanitic, QAPF diagram, the pyroclastic diagram, mafic diagram, ultramafic diagram, or lamprophyre.

That is all that is required for identifying and igneous rock via the petrographic analysis outlined in this book. You can go much deeper. In addition you can include a chart of the test results you used to identify the minerals in the rock, like Mohs hardness, or streak tests. See p.98-100 for an example of naming an igneous rock.

Example of Naming an Igneous Rock

Below is a to scale photo I took of a slightly weathered rock in the field.

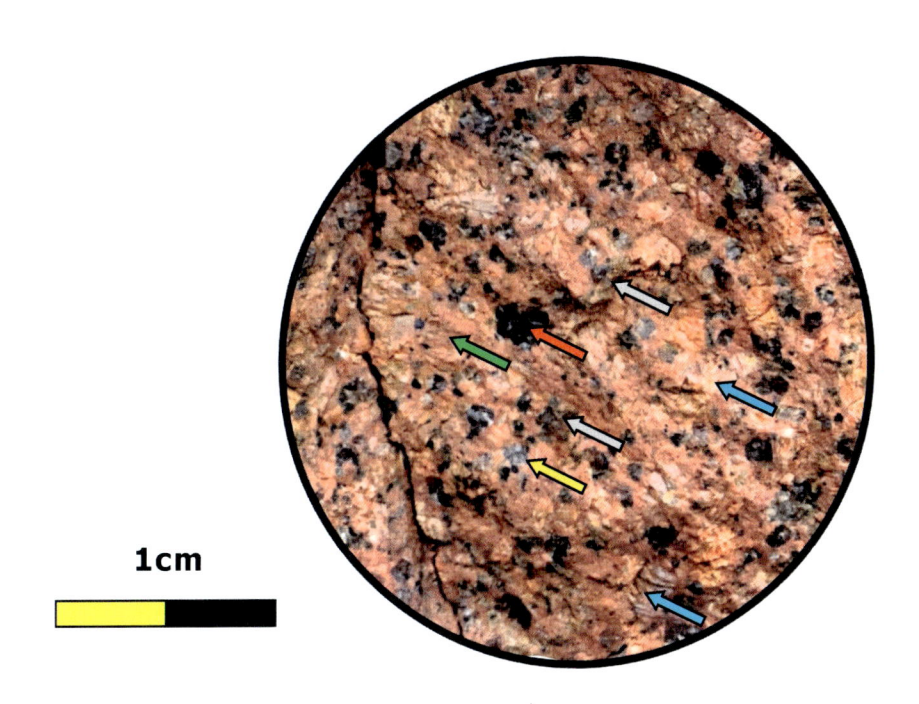

1cm

Minerals present

4% gray granular translucent medium grained quartz (yellow arrow). 8% pink and light yellow orange medium to coarse grained plagioclase (blue arrows). 11% black medium to coarse grained hornblende and biotite (red arrow). 7% light brown to light olive weathered out, medium to coarse grained pyroxene (gray arrows). 70% pale orange brown coarse grained alkali feldspar.

Based on the above information gathered, we know we have a generally medium to coarse grained phaneritic rock. So we will use the plutonic QAPF plot. First, lets normalize everything.

Normalized numbers used for plotting on the phaneritic QAPF diagram

Mafics = 18% (not plotted), Q = 5%, P = 11%, and A = 89%

Name

Now with my quartz at 5% this puts me right at the syenite-quartz syenite line. However, before rounding my normalized Q is ~4.88%, which is <5%, so I will go with syenite. I have 11% hornblende and biotite (which is less than but close to 15% and the crystals are prevalent) so I should include that in the name. I only have 7% pyroxene, so I will leave that out. The major color is in the alkali feldspar, so I will use that color. Since most of my alkali feldspar is coarse, I should include that in the name, although it's not required. So my most simplified rock name is:

> ***Light orange brown coarse SYENITE with HORNBLENDE and BIOTITE***

A more detailed name could be:

> ***Light orange brown coarse SYENITE, with black medium-coarse HORNBLENDE and BIOTITE, minor light olive medium-coarse grained PYROXENE***

That would be about as detailed of a rock name you should realistically get. However, your description can be more detailed.

WORKSHEET FOR QAPF PLOT

Below is a blank worksheet to fill out as you go in order to assist you in the visual rock classification process of igneous rocks. Copy as many as you need.

Worksheet for QAPF Plot

Sample ID #:		Date Collected:
Collected by:	Analyzed by:	
Field book name of original sample log:		Date analyzed:
		Page:

Aphanitic ☐   Porphyritic ☐

Mineral	Color	Average grain size	Raw %	1st normalized %	2nd normalized %

Mafic | Accessory minerals

Accessory minerals are removed from the volume of the rock, just as the mafics are.

Quartz		
Alkali feldspar		
Plagioclase		
Foids		

QAPF Plot (taken directly from where plotted on the diagram):

NOTES

Igneous Rock Samples Under High Magnification:

This rock is a nepheline rich rock from the Coldwell Complex in Ontario.

Diameter ~1 mm

The black nearly euhedral mineral at the center is hornblende. The subhedral black at the very top is biotite. The glossy luster is nepheline, NOT quartz.

Diameter ~1 mm

The red anhedral ring is K-spar. The more euhedral white crystal within the red ring is plagioclase, you can clearly see the striations. The black specks scattered within the plagioclase is likely pyroxene. The light mineral outside the red ring is nepheline. The black outside the ring is biotite.

Igneous Rock Samples Under High Magnification:

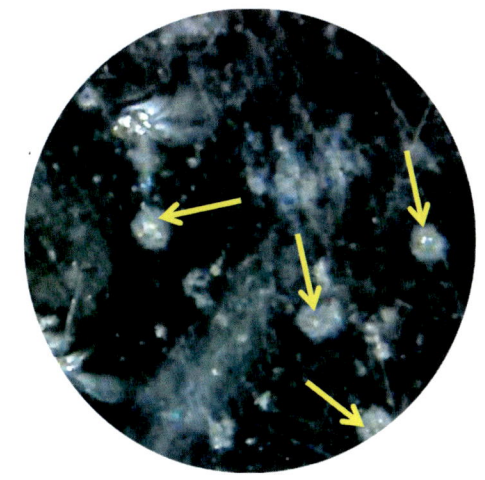

This is a piece of volcanic glass, likely obsidian. The diameter of the circle is ~1 mm. The well developed nearly white euhedral crystals are likely plagioclase (yellow arrows). The rest of the light colors is the mineraloid version of quartz. There is no crystal development. The mineraloids that make up the black could not be determined.

This is a Precambrian welded tuff from the Michigamme Formation near L'Anse in the Upper Peninsula of Michigan. The diameter of the circle is ~3.5 mm. This is a mixing of long angular and rounded grains. Some is quartz and volcanic glass. However, most of the rock is lithic fragments of microcrystalline grains, dark volcanics, and pale muddy matrix. There is also abundant feldspar. The grainsize varies 0.05-1.7mm in size.

Igneous Rock Samples Under High Magnification:

This is a phonolite with large white phenocrysts. The diameter of the circle is ~3.5 mm. This rock is not from the Midwest but it's a good example of phenocrysts in a dark aphanitic groundmass. The large transparent white crystals are mostly microcline and orthoclase. The darker parts is the feldspathoid naphthalene. There's minor augite, hornblende, and plagioclase.

This is native copper from the Upper Peninsula of Michigan. The diameter of the circle is ~3.5mm. Upon magnification the copper loses its orange color. The undulating shapes are typical of native metals. There are some black iron impurities and the "gold" colored parts are chalcopyrite. The greenish blues are other copper minerals due to weathering.

This page is intentionally left blank

Lab Reference Book
Classifying Metamorphic Rocks

Steven D.J. Baumann, PG

See p. 137 for rock description on previous page

INTRODUCTION

Rocks collected in the field are often brought back for more detailed analysis in a lab setting. The identification of rocks based off small scale texture and mineral composition is petrology. Identifying the composition of a rock is just as important as identifying its large scale structures in the field.

This book is a general guide for identifying metamorphic rocks visually in the lab. What do I mean by visually? I mean either through simple tests or through visual inspection that does *NOT* require thin sections. Most of us post-school do not have easy access to academic institutions where we can cut thin sections. You can still use a hand lens, magnifying glass, or a microscope.

Herein, I deal with major rock forming minerals in metamorphic rocks and some common accessories. The primary focus of this book is on rocks common in the Midwest of North America. This book is not meant to be all inclusive, nor is it meant to include information for every possible rock collected from the field.

The classification of metamorphic rocks is very complex. Unfortunately, many involve some sort of origin assumption. I am trying to avoid that in this book. I tried to develop a strictly lithological classification system, or modify existing ones.

This is the last of three volumes. There is also an igneous (Baumann, 2019[2]) and a sedimentary (Baumann, 2019[3]) lab reference book. Many textbooks and classes on petrology will combine both igneous and metamorphic rocks. I am separating them. There is enough variety in metamorphic rocks to justify a separate book. Although the igneous rock book, with this one, would provide a more complete picture as a lot of the identification concepts are similar to one another.

All photos were taken by the author, unless noted otherwise. All drawings and diagrams are by the author. Adaptations are noted.

Important Notes When Using This Book

This book is designed to aid in basic petrologic analysis in a laboratory setting. It is not intended to be used if detailed petrographic analysis is being conducted. If you have the ability to do thin sections, use an electron microscope, or run chemical analysis, this book can only be used as a generic guide. As a result, some things are left out.

All the detailed mineral data in this book was taken from the online version of the "Handbook of Mineralogy" (see references under: Anthony, et al.).

I hope you find this book informative and helpful! It's based not only off research but actual experience over the years as I attempt to classify rocks in a standard manner.

PART I:
OVERVIEW AND GENERAL CHARACTERISTICS OF METAMORPHIC ROCKS

WHAT IS A METAMORPHIC ROCK?

This question isn't as easy to answer as it may seem. The definition used herein is: "a rock that has undergone alteration and possible from heat and pressure reflected in the rock as some degree of ductile deformation, but not enough heat and pressure to melt the entire rock" (adapted from: Vernon and Clarke, 2008). The process can start as early as 100°C (212°F) at 1 gigapascal (GPa), or 0.1 kilobars (kb) of pressure, or ~9869.2 atmospheres of pressure. This corresponds to a depth of ~3.3km (2.05 miles) on stable continental or oceanic crust (not in subduction or spreading zones).

Even with this ridged sounding definition, there are still gray areas. Low grade metamorphism and diagenesis (a sedimentary rock term only), do overlap slightly (p. 119). The line between igneous and metamorphic can often be blurred. As the edge of granitic plutons will often exhibit banding structures similar to gneiss as the cooling magma flowed. In this situation, there is no real way to differentiate between a granite and a gneiss from a rock in a box. The best thing to do is look to see if the crystals are deformed, see if they show signs of strain. If yes, then it's a gneiss; if no, then it's a granite.

Another common gray area isn't with the rocks but is with the person identifying them. People in different regions may describe things differently. This is most common when the protolith (parent rock) is sedimentary. The Baraboo Formation is mostly a quartzite that experienced low to medium grade metamorphism. Most geologists south of the 45th parallel, would have no problem calling it a quartzite. If you go north to Canada in the Lake Superior and Lake Huron areas, meta-sedimentary rocks that experienced greater metamorphism than the Baraboo Formation did, are often called by their sedimentary protolith names (Casshyap, 1968). For example: sandstone instead of quartzite, shale instead of slate or phyllite, and dolomite instead of marble. I don't agree with this treatment. If a rock has clearly experienced some metamorphism, it should be called by its metamorphic name.

It is also important to note that chemical alteration and extreme hardness caused by chemical cementation (e.g. calcite and hematite) often in sandstones and limestones, does NOT qualify as metamorphism. Those fall under lithification and diagenesis.

METAMORPHIC ROCK TYPES

The below chart depicts the root name for most metamorphic rocks. Almost all metamorphic rocks can either be fit into foliated or non-foliated. Ironite would technically be a non-foliated rock derived from banded iron formation (BIF) since the bands are primary sedimentary structures and not part of the metamorphic process. Migmatite (p. 151-153) and skarn (p. 154-155) are a special cases and are not depicted below.

Quartzite and marble are extremely stable under most metamorphic regimes, so they tend to remain as those rock types throughout the process. Other sedimentary rocks such as mudstones, proto-quartzite, immature quartzite, conglomerate, and even ironite will eventually become schists and gneisses under higher grades of metamorphism.

Rock Name	Texture	Grain Size (Wentworth)	Diagnostic Characteristics	Protolith (Parent Rock)
SLATE	Foliated	Subcrystalline (lutite)	Well developed cleavage with smooth dull surface	Mudstone
PHYLLITE	Foliated	Subcrystalline to fine crystalline (lutite-pelite)	Breaks along uneven surfaces, glossy	Slate, rarely sandite
SCHIST	Foliated	Fine to medium crystalline (psammite)	Mica is prevalent, scaly foliation	Phyllite, sandite
GNEISS	Foliated	Medium crystalline to macrocrystalline (psammite-psephite)	Compositional mineral segregation in bands	Schist, plutonic and volcanic rocks, sandite
MARBLE	Non-foliated	Microcrystalline to macrocrystalline (psammite-psephite)	Interlocking calcite/dolomite grains	Carbonates
QUARTZITE	Non-foliated	Fine to coarse crystalline (psammite)	Fused quartz grains, recrystallization of minor minerals	Quartz sandstone
PROTO-QUARTZITE	Non-foliated	Fine to coarse crystalline (psammite)	Fused grains, recrystallization of minor minerals	Feldspar rich sandstone
IMMATURE QUARTZITE	Non-foliated	Fine to coarse crystalline (psammite)	Fused grains, recrystallization of minor minerals	Lithic rich sandstone
CONGLOMERITE	Non-foliated	Variable	Fused grains, recrystallization of minor minerals, breaks through grains not around	Conglomerate
IRONITE	Non-foliated	Microcrystalline to coarse crystalline (lutite-psammite)	Fused grains, interlocking iron mineral grains, recrystallization of minor minerals, boudinage common	Banded iron formation (BIF)

Adapted from: Peter Mulroy's earth science page at peter-mulroy.squarespace.com/how-do-we-identify-metamorphic-rocks

Anhedral and euhedral are terms restricted to describing igneous and metamorphic rocks.

For field identification purposes, anhedral minerals have no defined crystal facies. Euhedral minerals have well developed crystal facies and subhedral is in between. Rocks rarely exhibit all minerals within as euhedral. Usually only one or two minerals will be euhedral and the rest will be subhedral to anhedral. Euhedral crystals tend to be the first that form from alteration in a metamorphic rock.

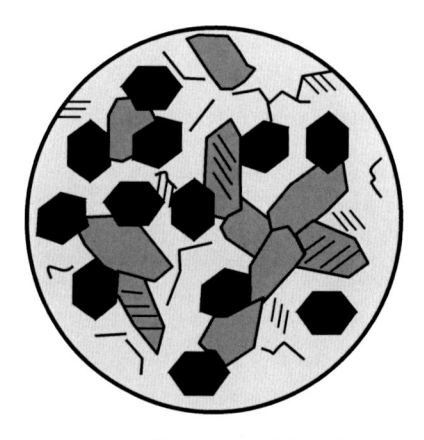

Euhedral, Subhedral, and Anhedral

Here the dark crystals are euhedral while the gray are subhedral and the light gray are anhedral.

Equigranular is a term reserved for igneous/metamorphic rocks that means 80%+ of the crystals are of nearly the same magnitude or size.

GRAIN SHAPE OF CRYSTALLINE METAMORPHIC ROCK

Crystalline metamorphic rocks bear minerals that may have crystalline facies or granular facies. There are two main types; idioblastic and xenoblastic. **Idioblastic** are bound by their crystal facies (Nelson, 2018), they are crystalline. **Xenoblastic** ones have little to no discernable crystal facies (Nelson, 2018). They are granular. For the sake of using as little technical terms as possible, crystalline and granular are used in this book.

These two traits together form the **crystalloblastic** series. What that means is a mineral tends to be more euhedral when physically against one lower in the series. Although this holds true for all igneous rocks cooled from melt, due to Bowen's Reaction Series, it also applies to metamorphic rocks that have undergone recrystallization.

Crystalloblastic Series

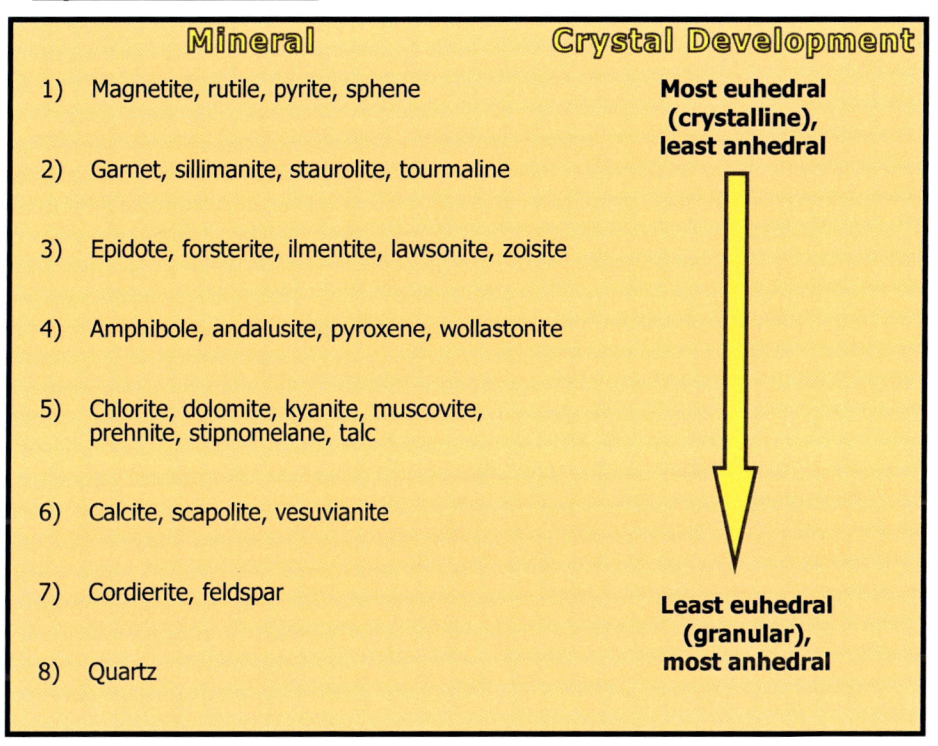

	Mineral	Crystal Development
1)	Magnetite, rutile, pyrite, sphene	Most euhedral (crystalline), least anhedral
2)	Garnet, sillimanite, staurolite, tourmaline	
3)	Epidote, forsterite, ilmentite, lawsonite, zoisite	
4)	Amphibole, andalusite, pyroxene, wollastonite	
5)	Chlorite, dolomite, kyanite, muscovite, prehnite, stipnomelane, talc	
6)	Calcite, scapolite, vesuvianite	
7)	Cordierite, feldspar	Least euhedral (granular), most anhedral
8)	Quartz	

Adapted from: Winter (2001)

CRYSTALLINE SIZE

Crystalline (sometimes called grain size in casual speak) sizes for metamorphic rocks are on the next page and are a modification of the Wentworth scale.

There are two basic terms used. A derivation of the word "Crystalline" for metamorphic rocks with crystalline structure; e.g. microcrystalline, subcrystalline, crystalline, and macrocrystalline.

The second division, in parenthesis on the next page, are for non-crystalline granular metamorphic rocks; e.g. Lutite, pelite, psammite, and psephite. These terms are borrowed from sedimentary terms but aren't really used anymore. So herein they are adapted to metamorphic rocks.

The term lutite is only to be applied if the metamorphic rock shows an overall granular texture. If a combination between granular and crystalline is encountered in the rock, the term microcrystalline should be used.

Adapted Wentworth Scale Chart for Crystalline Size

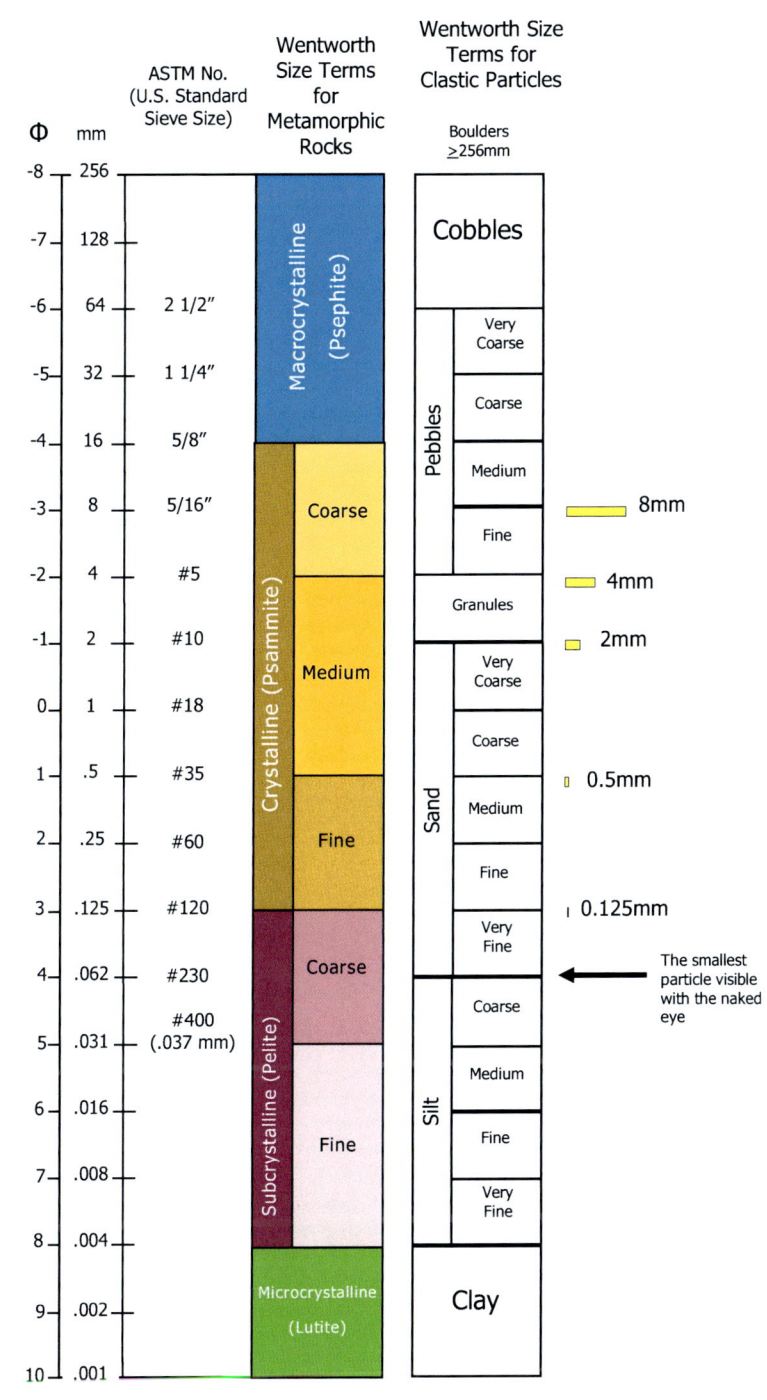

Φ	mm	ASTM No. (U.S. Standard Sieve Size)	Wentworth Size Terms for Metamorphic Rocks	Wentworth Size Terms for Clastic Particles		
-8	256		Macrocrystalline (Psephite)	Boulders ≥256mm		
-7	128			Cobbles		
-6	64	2 1/2"		Pebbles	Very Coarse	
-5	32	1 1/4"			Coarse	
-4	16	5/8"			Medium	
-3	8	5/16"	Coarse		Fine	8mm
-2	4	#5	Crystalline (Psammite)	Granules		4mm
-1	2	#10	Medium	Sand	Very Coarse	2mm
0	1	#18			Coarse	
1	.5	#35			Medium	0.5mm
2	.25	#60	Fine		Fine	
3	.125	#120			Very Fine	0.125mm
4	.062	#230	Coarse	Silt	Coarse	← The smallest particle visible with the naked eye
5	.031	#400 (.037 mm)	Subcrystalline (Pelite)		Medium	
6	.016		Fine		Fine	
7	.008				Very Fine	
8	.004			Clay		
9	.002		Microcrystalline (Lutite)			
10	.001					

Adapted from: Wentworth (1922) and Baumann (2019 [2,3])

TYPES AND CAUSES OF METAMORPHISM

There are essentially 6 metamorphic situations on Earth. The chart below summarized them. There can be more than one metamorphic type within the same area.

Type of Metamorphism	Cause of Metamorphism	Characteristics of Metamorphism	Extent of Metamorphism
Burial (or load) *a non-tectonic form of regional metamorphism*	Deep burial in a sedimentary basin or passive margin. Dominantly pressure metamorphism.	Deepest sediments are metamorphosed. The shallower ones typically are not.	Variable on the depth and surface area of the basin/passive margin in which it formed.
Cataclastic	Mechanical deformation caused by friction heating and pressure.	Low to high grade metamorphism in small, near human scale lenses or sheets.	Localized to the fault or shear zone in which the metamorphism occurred.
Contact	Igneous intrusion into host rock causing metamorphism through increased temperature and limited increased pressure.	Degree of metamorphism is greater towards the igneous intrusion. Away from the intrusion it decreases. The area of metamorphism is known as the "metamorphic aureole".	Depends on the size and temperature of the igneous body. The hotter and larger the magma, the greater the metamorphic aureole. Can be on the scale of meters to about 3km (~2 miles) from the intrusion.
Hydrothermal *a form of non-melt induced contact metamorphism*	Heated water, steam, and other fluids intruding into host rock, causing metamorphism through increased temperature.	Metamorphism is greater to the hydrothermal source but generally limited. Ores are often found in such deposits. It is low grade metamorphism.	Usually very localized to fractures that the fluids moved through.
Regional	A large and prolonged, and usually compressive tectonic event; such as an orogeny. Always increased pressure. Increase heat if the rock is buried deep enough.	Usually only low to medium grade outside of localized intense metamorphism. Cataclastic, contact, and/or hydrothermal metamorphism is common within regional metamorphism.	Large areas, usually >1000km² in area where all or near all the rocks show some degree of metamorphism.
Shock	A single catastrophic event. An extraterrestrial body impact or an intense volcanic eruption. Very high pressure metamorphism.	"Shocked quartz" is often produced such as coesite and stishovite. Partial melting will occur at ground zero and may occur if the crust is thin or heated.	Variable. It depends on the size and depth of the crater and/or volcano.

Adapted from: Nelson (2018)

METAMORPHIC FACIES

Metamorphic facies are not part of the name but can be included in detailed descriptions, and they can aid in interpreting the evolution of a metamorphic rock.

Below is an alteration of the chart on the next page depicting the relative "grades" of metamorphism. Both the chart below and on the next page follow shales from diagenesis to high grade metamorphism (slate, phyllite, schist, gneiss, and migmatite). Remember "diagenesis and lithification" are terms for sedimentary rocks *NOT* igneous and metamorphic rocks. Almost any clastic, plutonic, or volcanic rock that undergoes high enough grades of metamorphism will eventually become a schist, gneiss, and migmatite. The purpose of separating the charts is to minimize the amount stuff clustered on to it.

Lithification and diagenesis are not technically metamorphic grades. Nor are they consistent across rock types, that's why diagenesis overlaps with zeolite facies a bit. They are just included for reference and scale to illustrate what occurs before metamorphism begins. They both overlap hydrothermal, but hydrothermal isn't relevant unless hot water is injected from below.

Metamorphic Grade Temperature and Pressure Diagram

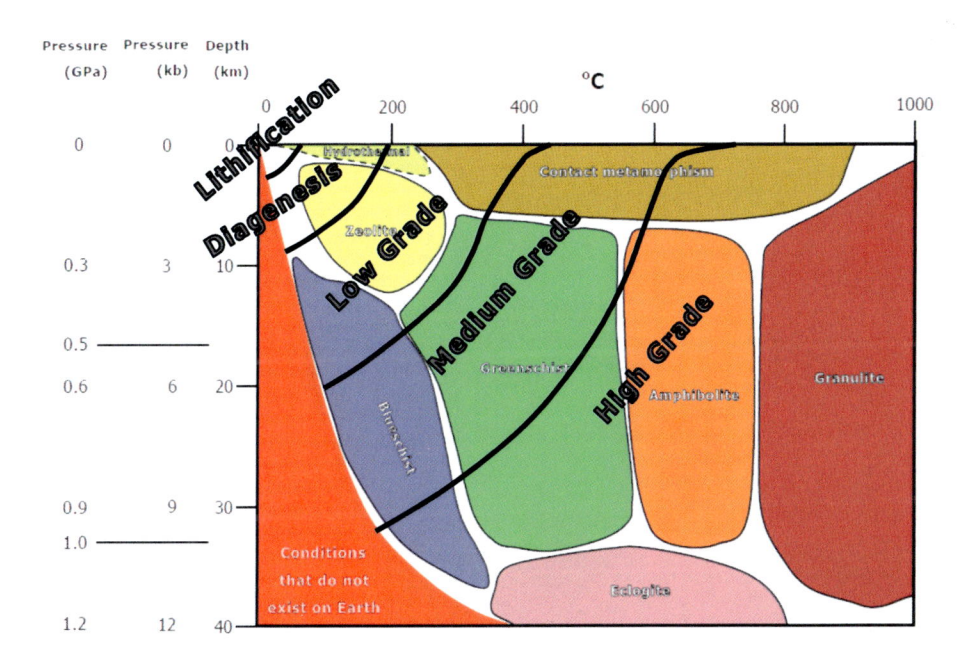

Adapted from: Robertson (1999) and Nelson (2018)

Metamorphic Facies Temperature and Pressure Diagram

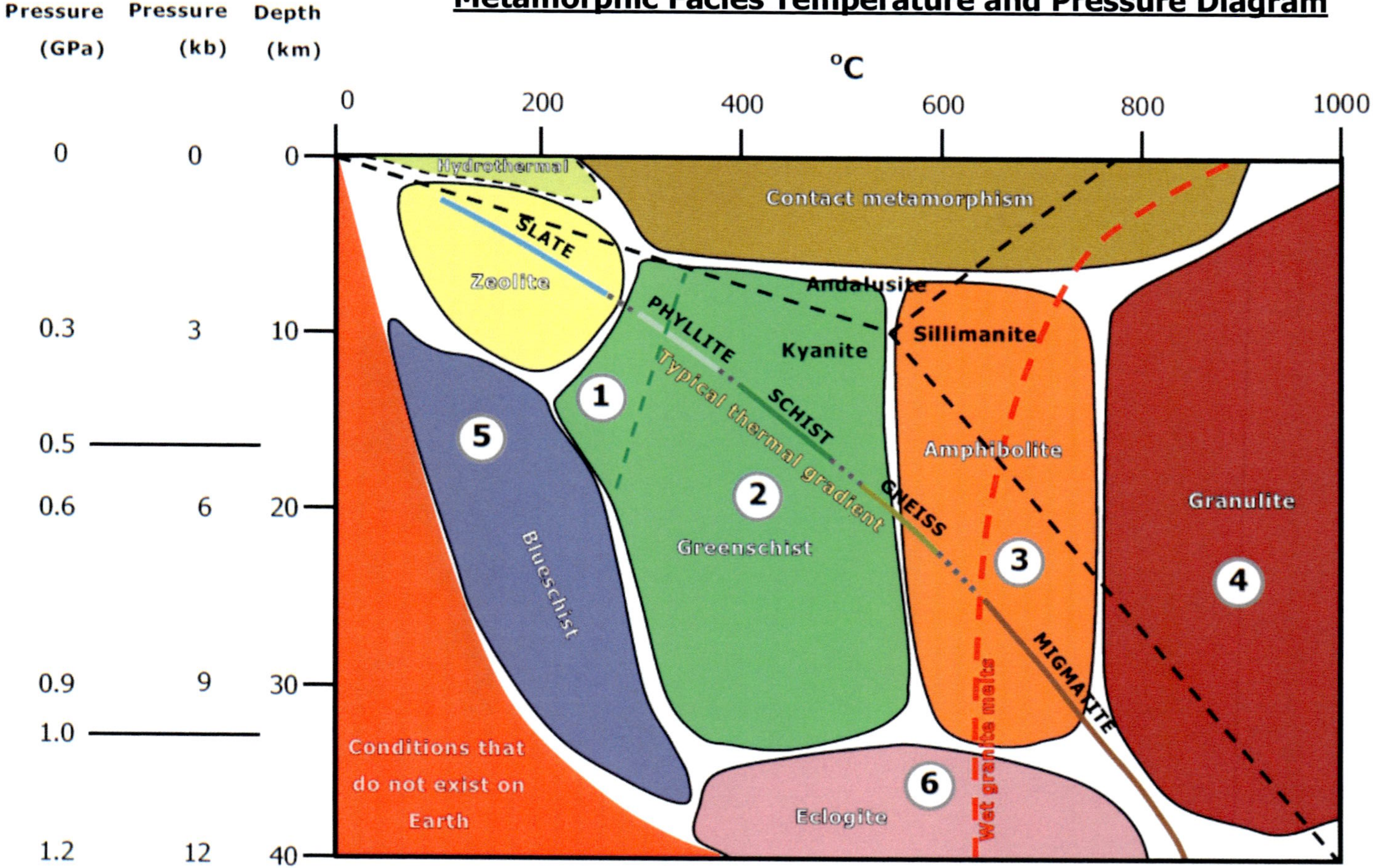

Adapted from: Robertson (1999)

Legend (for Metamorphic Facies Temperature and Pressure Diagram)

This is the legend for the previous page. The aluminosilicates "andalusite, kyanite, and sillimanite" zones indicate that those minerals are stable at those temperatures and pressure. If those minerals are present, it indicates a framework for how deep and how hot the rock got.

The curve leading from slate to migmatite, it a "normal geothermal gradient curve" for mudstones. The gray dotted lines between rock types in the "normal geothermal gradient" are areas of overlap by the neighboring rock types. The depicted "normal gradient curve" assumes stable continental or oceanic crust. The curve raises for contact metamorphism and lowers in subduction zones.

The fixed point (where kyanite, andalusite, and sillimanite meet and can all exist in a rock) is at about 3-5kb at a temperature of 520°-580°C. The reason why this is not an exact fixed point (as the name implies), is because of the varying mineralogy percentages in the protolith; although the maximum is often considered the fixed point at 5kb and 580°C (Vernon and Clarke, 2008).

Mineral assemblages

The numbers are minerals that can be expected in those zones. <u>NOTE:</u> The area of number 1 (left of the green dashed line in greenschist, is sometimes divided out as the Prehnite-pumpellyite facies (Robertson, 1999). Most charts do not included it and I consider it splitting hairs. The term "lower greenschist" can be used. However, it is included here because it frequently appears in the literature like in Robertson (1999). The below assemblage are for metamorphosed mudstones.

1) Laumontite, prehnite + pumpellyite, prehnite + actinolite, pumpellyite + actinolite, pyrophyllite

2) Actinolite + chlorite + epidote +albite

3) Hornblende + plagioclase, staurolite

4) Orthopyroxene + clinopyroxene + plagioclase, sapphirine, osumilite, kornerupine. *NO* staurolite, *NO* muscovite

5) Glaucophane, lawsonite, jadetic pyroxene, aragonite. NO biotite

6) Omphacite + garnet. *NO* plagioclase

Adapted from: Robertson (1999)

COMMON METAMORPHIC MINERALS

Remember, when running a hardness test, you have to make sure that you are pressing into the mineral only. Do not cross over to other minerals. If crystals aren't large enough to be scraped, then you cannot perform a hardness test. When conducting the test, use a significant bit of pressure. Don't just lightly drag the tool over the mineral. That being said, don't press so hard that you are trying to dig in, you could break the tool. Below and on the next page, are charts of common mineral characteristics in metamorphic rocks. This list is in no way comprehensive.

MINERAL	HARDNESS RANGE	STREAK	OTHER CHARACTERISTICS
Actinolite	5-6	White	Bladed but more commonly a fibrous amphibole. Fibrous forms are elastic. Usually distinctive green color, but can be yellowish to nearly white.
Albite	6-6.5	White	Striations. It's a plagioclase feldspar.
Andalusite	6.5-7.5	White	Trimorphic with kyanite and sillimanite.
Ankerite	3.5-4	White	Carbonate rock. It is best described as an iron bearing dolomite and resembles it closely in appearance, even belonging to the same crystal system. It's much denser a 2.93 to 3.12g/cm^3. Reacts weakly with dilute HCl.
Aragonite	3.5-4	White	An unstable carbonate mineral. Highly reactive with weak HCl.
Augite (is a clinopyroxene)	5.5-6	White with a greenish hue	SG = 3.19-3.56, very vitreous luster, can be prismatic but often is dendritic. Stable at lower temps unlike the closely related pigeonite.

MINERAL	HARDNESS RANGE	STREAK	OTHER CHARACTERISTICS
Biotite	2.5-3	White	Occurs as angular lithic mica flakes in clastic rocks.
Calcite	3	White	Reacts strongly with cold dilute HCl.
Chlorite	2-2.5	Pale green to light gray	Usually various shades of gray. Chlorite is a group of minerals. The most common chlorite minerals are chamosite, clinochlore, and pennantite. Flexible but not elastic.
Cummingtonite	5-6	Very light gray to white	Forms bladed, columnar, or fibrous crystals. Common in amphibole/magnesium rich metamorphic rocks
Copper (native element)	2.5-3	Shinny orange red	————
Diopside (is a clinopyroxene)	5.5-6.5	White	SG = 3.278, common in skarn and contact metamorphic environments, distinct and very good cleavage in one direction when short tabular crystals are present. It tends to be massive or granular.
Dolomite	3.5-4	White	Reacts weakly with cold dilute HCl
Enstatite (is an orthopyroxene)	5-6	Gray	SG = 3.2-3.3, perfect cleavage in two directions at 90° to each other. Crystals are rare but tend to be laminar, more commonly fibrous or massive.
Epidote	6-7	Very light gray to white	SG = 3.3-3.6, tends to be yellow green to a light green gray, prismatic with striations, perfect cleavage in one direction imperfect in one direction
Fayalite (olivine group)	6.5-7	White	SG = 4.392, can react with oxygen to form magnetite. Green yellow to yellow brown in color.
Feldspars (general)	6-6.5	White, rarely light gray	Will often appear red and yellow in sandstones. Striations on grains are common.

MINERAL	HARDNESS RANGE	STREAK	OTHER CHARACTERISTICS
Garnet Group (almandine is the most common in metamorphic rocks)	6.5-7.5	White	Rhombic or dodecahedral isometric crystals, some minerals fluoresce (almandine does not, grossular will very faintly yellow-orange in both short and long wave UV), tends to be a translucent to opaque deep red to purple color.
Glaucophane (amphibole)	6-6.5	Blue gray	Crystals tend to be long and slender, usually massive to granular. Blueschist metamorphic facies. Tends to be very blue in color.
Goethite (a.k.a. brown ocre)	5-5.5	Brown to yellow brown	Very weakly magnetic.
Graphite	1-2	Black	SG = 1.9-2.3, feels greasy to the tough, metallic to earth luster, almost always microcrystalline. Forms from a reduction in sedimentary carbon rocks during metamorphism.
Grunerite	5-6	Colorless	SG = 3.45, most commonly forms columnar crystals but can be fibrous or massive. Color is commonly light gray brown.
Hematite	5-6	Bright red to bright dark red	Most common iron oxide in BIF. Magnetic if heated to the Curie Point (769.9°C).
Hornblende	5-6	Very light gray	SG = 2.9-3.5, opaque, color is very dark green to black. Imperfect cleavage at 54o and 124o. Can be granular but tends to form nice hexagonal crystals, since it is one of the first minerals to crystalize in Bowen's reaction series. Common in gneiss and schist.
Kornerupine	6-7	White	SG = 3.29-3.35, vitreous luster. Crystals are rare and tend to be prisms. Massive to fibrous.

MINERAL	HARDNESS RANGE	STREAK	OTHER CHARACTERISTICS
Kyanite	4.5-5 in one axis, 6.5-7 in the perpendicular axis	White	SG = 3.53-3.67, tends to be bright blue in color. Tends to be columnar or bladed, sometimes fibrous. Principally forms in gneiss and schist (and even migmatite) derived from mudrocks. Regional metamorphic.
Jadeite (is a clinopyroxene)	6.5-7	White	SG = 3.24-3.43, color tends to be a bright green and occasionally light bright purple. Bright colored versions will fluoresce in short or long wave UV with high color variance.
Laumontite	4	White	Very common in metamorphic rocks.
Lawsonite	7.5-8	White	SG = 3.09, it forms in 6-12kb pressure and 300-400°C metamorphic environments. Vitreous luster. Brittle fracture. Cleavage is perfect in two directions and imperfect in one.
Magnetite	5.5-6	Black	Magnetic and only stable in low grade metamorphic environments.
Olivine (olivine group)	6.5-7	None	SG = 3.2-4.5, typical color is olive green. Vitreous luster. Conchoidal fracture. Tends to be massive to granular. Stable into the mantle.
Omphacite (is a clinopyroxene)	5-6	White with a greenish hue	SG = 3.16-3.43, forms anhedral crystals or exists as granular or massive. Usually indicates blueschist facies with epidote, glaucophane, lawsonite, and titanite.
Osumilite	5-6	Blue gray	SG = 2.62-2.63, forms tabular to prismatic crystals but can be granular to massive. Forms in high grade metamorphic regimes.

MINERAL	HARDNESS RANGE	STREAK	OTHER CHARACTERISTICS
Pigeonite (is a clinopyroxene)	6	Very light gray to nearly white	SG = 3.17-3.46, generally unstable (depending on relative amounts of Ca, Mg, and Fe) <900°C and the related mineral augite tends to be present in the lower temperatures.
Prehnite	6-6.5	White	SG = 2.8-2.95, tends to be very brittle. Pearly luster. Generally colorless to gray with slight hues of green or yellow. Very low end of the greenschist facies.
Pumpellyite	5.5	White	Density = 3.2g/cm^3, it tends to be light green to olive green. Forms nice interlocking circular radiating disks composed of fibrous to laminar crystals. Very low end of the greenschist facies.
pyrophyllite	1.5-2	White	SG = 2.65-2.9, tends to be nearly white in color to light greens or browns. There are no known macro crystals but its habit is similar to pumpellyite forming disks. It can also be massive or granular.
Quartz	7	White	SG = 2.65 in pure samples.
Riebeckite	6	Pale blue to bluish gray	SG = 3.28-3.44, tends to be very dark blue to nearly black in color. Occurs as prismatic crystals or it can be earthy, massive, or very fibrous. Common in Ironites. Stable up to amphibolite facies.
Sillimanite	7	White	SG = 3.24, forms prismatic or needle like crystals or it can be fibrous. Stable in the low-high pressure and high temperature of the amphibolite and granulite facies.

MINERAL	HARDNESS RANGE	STREAK	OTHER CHARACTERISTICS
Staurolite	7-7.5	White to light gray	Usually prismatic and commonly twins at 60° and 90°. Stable in medium to high grades of metamorphism in a regional regime. Commonly occurs with, almandine, biotite, and kyanite in schist and gneiss.

HORNFELS AND GRANOFELS PROBLEM

Honfels are not a problem so much as they are creating unnecessary complexity. Hornfels are basically non-foliated granoblastic rock in the immediate baked zone of an intrusion (Vernon and Clarke, 2008). The term is sort of redundant, and implies an origin more than being a descriptive term. The term "baked zone" in universally recognized. A meta-sedimentary rock that is a hornfel can just be described as baked.

Granofels are just granular metamorphic rocks which hornfels are a part of and contain mostly quartz and feldspar (Robertson, 1999). Since both foliated and non-foliated rocks can contain granular clasts, the term is redundant and somewhat confusing. It could be used as a waste bin term for granular rocks that can't be easily classified as schists and gneiss. The protolith of many granofels is usually proto-quartzite.

MYLONITE PROBLEM

Mylonite is both a rock type and a texture. Mylonite is basically defined as a "tightly banded mafic metamorphic rock that is fine crystalline and contains highly deformed clasts of other felsic rock within" (adapted from: alexstrekeisen.it/english/meta/myloniticmarble.php). Others take to a definition implying an origin. Such as a "metamorphic rock displaying brecciated schistosity, often associated with fault or shear zones due to cataclasis" (Allaby, 2008 and Kurniawan et al., 2009).

In geology we will often use words as both nouns and adjectives, like granite and granitic. In the case of granite, there isn't better rock name. Mylonite is different. There is no consistent definition of the word. I do not view it as an actual rock type; mainly due to so many definitions implying an origin of the rock. Even a descriptive definition (first definition above) it can better be described as a mafic gneiss or schist. The term "mylonitic" is often used as an adjective, for foliated rocks, placing it before gneiss and schist (Higgins, 1971). Plus it is not always related to faults and shear zones. It is also found in migmatite (see next page).

Herein, mylonite is not used as a noun. Mylonitic is used as an adjective for gneiss and schist. Its definition is as follows:

Mylonitic: A metamorphic rock that is a subcrystalline to fine crystalline mafic rock, that exhibits a fabric with tight bands and a consistent preferred orientation. It contains within it, deformed clasts of either host rock (if in a fault or shear zone) or recrystallized or igneous rock (if included as the metamorphic component in migmatite).

Words like protomylonite, ultramylonite, and blastomyllonite (Higgins, 1971) are abandoned herein.

Mylonitic texture in a road cut. It is composed mostly of mafic amphibolite in Ontario, Canada. GPS: 48.764°N, 85.430°W. The yellow arrows are pointing to a couple of the deformed felsic tonalite. Canadian dollar coin for scale. This rock has a small fault in it, but it is in a migmatite setting, not a shear zone (Baumann, 2016).

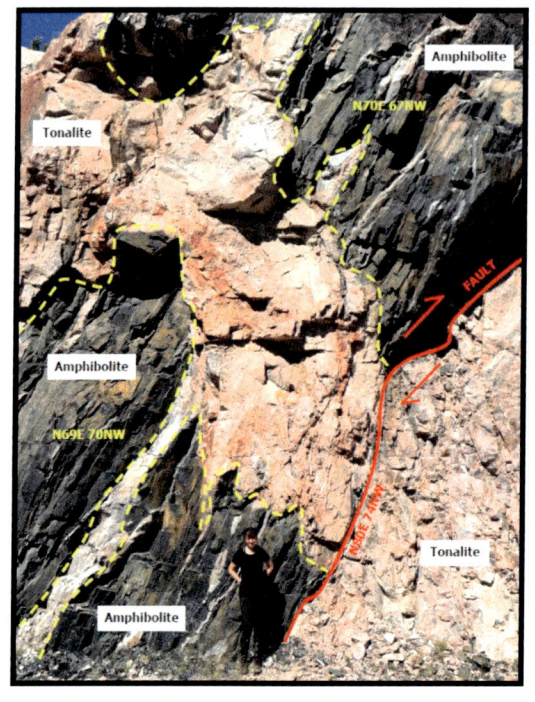

Same outcrop as above, with a person for scale.

This shows the metamorphic mylonitic amphibole gneiss (labeled as amphibole) crosscut by the igneous pink felsic tonalite; thus making the overall rock an "Mylonitic ARTERITE MIGMATITE". The two rock types are separated by yellow dashed lines.

The yellow letters and numbers are strike and dip on the gneiss, notice how consistent it is.

The red line is a small fault that formed way after the rock did. Red arrows show relative movement.

CHARACTERIZING BEDDING, FOLIATIONS, LINEATIONS, AND BANDING

Bedding, foliations, lineations, and banding all have one thing in common; whether they are very thin, thin, medium, thick, or very thick. Other than that, the similarities stop. If bedding is preserved and the protolith is definitely sedimentary, use the sedimentary thicknesses for laminations and beds (Baumann, 2019[3]). For everything else, use the chart below followed by the appropriate word, e.g. foliated or lineated. Banded would also be acceptable in lieu of foliated or lineated.

Thickness in cm	Foliations/lineations (To Scale)	Thickness in inches
2.5 – 0	Very thick lineation/foliation	0.98
1 – 0	Thick lineation/foliation	0.39
≤0.4	Medium lineation/foliation	≤0.16
≤0.3	Thin lineation/foliation	≤0.12
≤0.2	Very thin lineation/foliation	≤0.08
≤0.1	Micro lineation/foliation	≤0.04

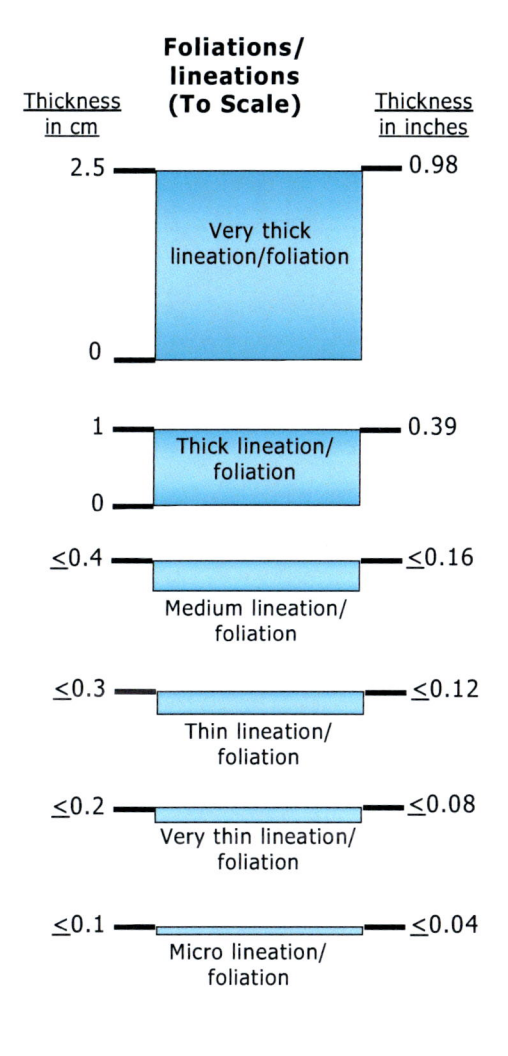

NOTE: *This scale is not to be used for recognizable primary structures in meta-sedimentary rocks; such as beds, cross beds, and laminations. The terms foliation and cleavage are often used interchangeably, when talking about slate.*

CLEAVAGE

Cleavage is a type of rock fabric that deals with how planar features develop in a rock caused by external stresses or initial deposition, and the rock's tendency to break easily along those developed planes (Leith, 1905). Fractures in rock are not splits along developed planar surfaces and can be in any other orientation. Although included herein with metamorphic rocks, it can occur in igneous and sedimentary rocks as well. Rock cleavage is not the exact same as mineral cleavage.

Types of Cleavage

Cleavage	Spaced	Compositional
		Disjunctive
		Crenulation
	Continuous or penetrative	Slaty (or fine) <0.01mm of spacing
		Platy (or coarse) >0.01mm of spacing

NOTE: *Compositional cleavage is cleavage that forms during deposition.*

Adapted from: Taylor (2008) structural geology lab; wou.edu/las/physci/taylor/es406_structure/

Graphic Depiction of Cleavage

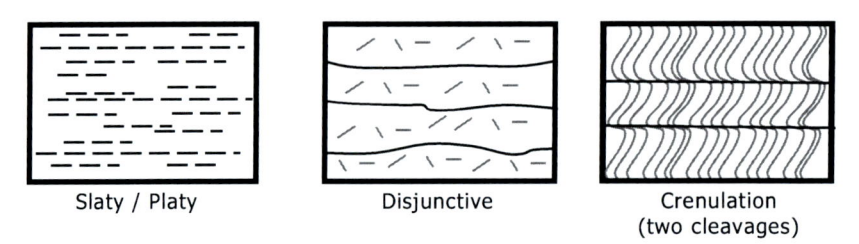

Slaty / Platy   Disjunctive   Crenulation (two cleavages)

NOTE: *A rock does not have to be a slate to express slaty cleavage.*

Adapted from: Winter (2001)

NAMING PROCESS

When describing a metamorphic rock you can be as detailed as you want. When naming one, you need to narrow it down. The following parts are a guide for a detailed but not over descriptive rock name. Their order is as listed below.

There are two *qualifiers* that are optional and only included in the name if a detailed description isn't given elsewhere. You have color, which is the **visual qualifier**. You have grain size , which is the **analytical qualifier**. These are lower case and separated by commas.

After the qualifiers comes the *modifying name*. There are also two types. There are textural modifiers and root modifiers. Some common **textural modifiers** are: Lineated, foliated, banded, brecciated, weathered, decayed, deformed, nodular, oolitic, flaggy, stromatolitic, mylonitic, cataclastic, ptygmatic, and boudinage. These are before a root modifier and only the first letter of each word is capitalized. Use as many as needed. They can also be used to describe one another, e.g. stromatolitic banded MARBLE. Describing the contact in certain rocks, like migmatite, can be useful; although not required. For example, is the contact sharp or gradational between distinct minerals.

Root modifiers are the second part of the modifying name. If the rock as a preferred characteristic similar to a closely related rock or has common attributes to two different rock types such as; gneissic, magmatic, phyllitic, and schistose. As many of these as needed can be used, and if applicable, they are not optional. They should be typed in italics and all capital letters.

The last and most crucial part is the **rock type** or **root name**. This is ALWAYS given and is usually listed in all capital letters. The only acceptable root names herein are; slate *phyllite, schist, gneiss, marble, quartzite, proto-quartzite, immature quartzite, conglomerate, ironite, skarn, arterite migmatite, and venite migmatite.*

At this point you are done for naming purposes.

PART II:
FOLIATED METAMORPHIC ROCKS

A foliated rock is a metamorphic rock that expresses either micro or macro banding caused by the alteration of the protolith during metamorphism. Most of the following definitions are adapted from Vernon and Clarke (2008).

SLATE

A slate is a low grade (usually zeolite facies) metamorphic rock whose protolith is a mudstone, volcanic ash, or a fine grained wacke.

It is differentiated from shale and siltstone by the presence of some aligned grains that point perpendicular to compressional stresses. Primary structures are intact.

Green slate of the upper Michigamme Formation looking in cross section at an outcrop on Lake Superior. Laminations are visible. The dark purple is sand from above.

PHYLITTE

A phyllite is low to medium grade (usually prehnite-pumpellyite to middle greenschist facies) metamorphic rock derived usually from a slate, but whose protolith can be a mudstone, volcanic ash, or a fine grained wacke.

Unlike slate, nearly all the grains show a preferred orientation perpendicular to compressional stresses. Primary structures are present to faintly preserved but discernable. Phyllite tends to have a slightly metallic luster to it.

Slate-phyllite of the Gowganda Formation looking down at a bedding plane with ripple marks.

SCHIST

A schist is a medium grade (usually upper greenschist facies) metamorphic rock derived from a multitude of different rocks, whose protolith is a mudstone, volcanic ash, non-quartz sandstone, granitic rock, and some volcanic rock.

Schist has nearly all the grains in a preferred orientation perpendicular to one or more stress fields. It also can involve a recrystallization of some minerals that can, but not always, be expressed as distinct light and dark bands. Primary structures either are or almost completely destroyed.

A cut piece of schist in cross section from Italy. Its protolith was an arkose.

GNEISS

A gneiss is a high grade (usually upper greenschist to middle amphibolite facies) metamorphic rock whose protolith is a mudstone, volcanic ash, non-quartz sandstone, granitic rock, and some volcanic rock.

Gneiss has all the grains in one or more preferred and distinct orientations perpendicular to one or more stress fields. It also can involve a recrystallization of some minerals that can that will be expressed as distinct light and dark bands. Primary structures are destroyed.

A gneiss (black and white bands) from a glacial erratic incorporated into a retaining wall in Madison Wisconsin. The light to deep red rock on the right is an intrusive dike.

PART III:
NON-FOLIATED METAMORPHIC ROCKS

A non-foliated rock is a metamorphic rock that does not usually express either micro or macro banding caused by the alteration of the protolith during metamorphism. These are usually chemically semi-stable rocks that can survive without much mineralogical alteration during metamorphism, at least in the low to medium grades. Some rocks can survive right up until melting as is the case with quartzite and marble. Most of the following definitions are adapted from Vernon and Clarke (2008) or Baumann (2019[1,2,3]).

MARBLE

A marble is a low to high grade metamorphic rock whose protolith is a limestone or dolostone with differing amounts of silica or other compounds. It can also be derived from carbonate volcanic rocks (but none exist in the Midwest).

It can be recognized from sedimentary carbonates by a fusing of the original grains or crystals.

Although marbles can withstand great heat and pressure, they are most commonly metamorphosed in the zeolite and low greenschist facies along with contact and hydrothermal metamorphism. Primary structures are usually intact although fossils can be deformed and can be mistaken for foliations.

CONGLOMERITE

A conglomerite is a low to high grade metamorphic rock whose protolith is a conglomerate. It can be described using sedimentary terms substituting "conglomerate" with "CONGLOMERITE".

Conglomerite can be recognized from sedimentary conglomerates by fractures running through the matrix and grains instead of around. Preferred orientation of pebbles is *NOT* necessarily an indication of metamorphism.

Conglomerites are most commonly metamorphosed in the zeolite and low greenschist facies along with contact and hydrothermal metamorphism. However, they can develop into schists and gneisses. Primary structures are usually intact up through low to middle grade metamorphism.

SANDITE

Sandite is metamorphosed sandstone (Baumann, 2019[1]). It is further broken down into quartzite, proto-quartzite, and immature quartzite. They are derived from an arenite protolith (see next page). There is no "wacke" version of them. The words SILTY and/or ARGILLACEOUS or MUDDY, would be put in front of the root name, in all caps.

Quartzite

Quartzite is the most common sandite in the Midwest; the Lorrain, Baraboo, and Sioux are extensive examples in the Midwest. It is a low to high metamorphic quartz arenite whose primary structures are usually preserved. Most Midwest quartzites fall in the zeolite or greenschist facies and are regional metamorphic rocks. There are some cataclastic quartzites like the Saint Peter Sandstone along the Sandwich Fault Zone in Illinois (p. 142). It can withstand high grade metamorphism.

Proto-quartzite

Proto-quartzite is spread out mostly in the lower half of the Huronian Supergroup of the Midwest, as well as red lenses within the Gowganda Formation. They are the result of regional metamorphism of arkosic sandstones. All the ones in the Midwest fall within the zeolite or lower greenschist (prehnite-pumpellyite) facies of metamorphism. They can theoretically exist intact up through the higher greenschist facies. After that, recrystallization would make it a schist or gneiss.

Immature quartzite

Immature quartzites are extremely rare. I have seen them in the Paleoproterozoic glacial deposits and in the Aweres Formation and scattered throughout the Huronian Supergroup. Some of the interflow sediments that occur in the Penokean orogeny will occasionally contain them as will graywacke. They are derived from the regional metamorphism of lithic sandstones. Due to containing a bag of different minerals and rock types, they wouldn't survive past medium greenschist facies. They would develop into schists and gneisses.

Sandite Classification Chart

The chart below functions as a normal ternary diagram. **NOTE:** *Feldsandite and lithsandite are not known to be present in the Midwest.*

(2nd order normalization)

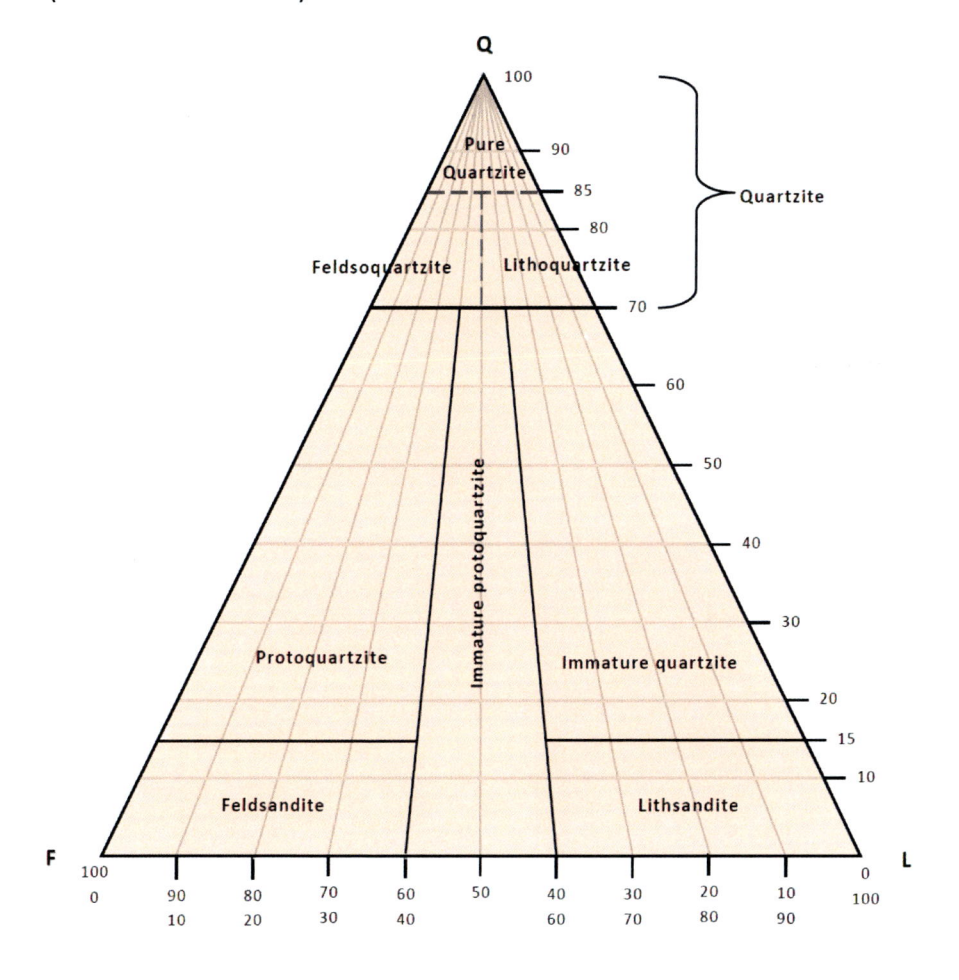

Lith is Greek for stone
Sandite is just a word to separate the metamorphic terms from the sedimentary ones.

Quartz includes monocrystalline and polycrystalline. Chert is put in with lithics and excluded from quartz as in Folk et al. (1970).

Specimens with less than 15% fines and matrix have nothing after the first word. If there is more than 15% up to 50±5% the word "wacke" is added to the end of the name.

Close-up (long length is ~1 inch (2.54cm) of a proto-quartzite, from a river pebble. Source is likely lower Huronian in origin.

An outcrop of conglomerite mostly (rounded white quartz and subrounded red jasper clasts) quartzite (white and pink host rock) of the Lorrain Formation on Ontario Route 638 in Ontario, Canada. Red Jasper is about 0.5-3.5 inches (~1.3 to 8.9cm).

Marble of the Gunflint Formation (gray host rock) with red and white seam agate. Near Thunder Bay, Ontario. ~50% actual size.

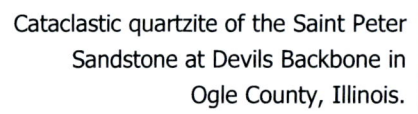

Cataclastic quartzite of the Saint Peter Sandstone at Devils Backbone in Ogle County, Illinois.

The photo below was taken by the author in August 2022. It was taken in Sillisit, Greenland.

The reddish rock is the quartzite of the Eriksfjord Formation. The black rock is the interbedded basalt lava flows and conglomerate. The rocks are Mesoproterozoic in age and are part of a failed arm of a rift.

CONGLOMERITE

Conglomerites are the metamorphic version of conglomerate (p. 192-193). The follow the same classification system except conglomerite is swapped for conglomerate. The conglomerites generally show clast deformation and break across clasts as where conglomerates break around clasts.

You can expand on you descriptions after naming the conglomerate. You can add things like the lithologies of the pebbles.

Paraconglomerate, Orthoconglomerate, and Dioconglomerate

Conglomerates come in three "main types". This will be the very last word of the name. "Paraconglomerate" is matrix supported; as is common with some diamictons and debris flows. "Orthoconglomerate" is clast supported; as is the case with river bars, some alluvial fans, outwash, and some desert pavement. "Dioconglomerate" are matrix and clast supported; as is common with bedded conglomerates like fluvial streams and some alluvial fans. Paraconglomerate, orthoconglomerate, and dioconglomerate will be your rock type.

Oligometic and Polyometic

There are two "subtypes" that can be added to the rock type you can choose, or leave off if you don't need a detailed description. "Oligometic" is for conglomerates who's clasts are $\geq$80% one rock type. "Polyometic" is used if there are clasts of <80% one rock type or multiple and varying lithologies. Subtype is always listed as the first part of the name.

Steven Baumann in August 2022 standing next to a conglomerite in southwest Greenland near Issormiut-Sillisit.

Conglomerites with Matrix

If a conglomerite has significant matrix (a.k.a. clasts finer than pebbles), the following chart is to be used and the appropriate modifier will be added in between the subtype and the type. The chart below is to be used for conglomerites with significant matrix, and is different from previous editions and is introduced herein. The chart gives you your matrix modifier if applicable.

(2nd order normalization)

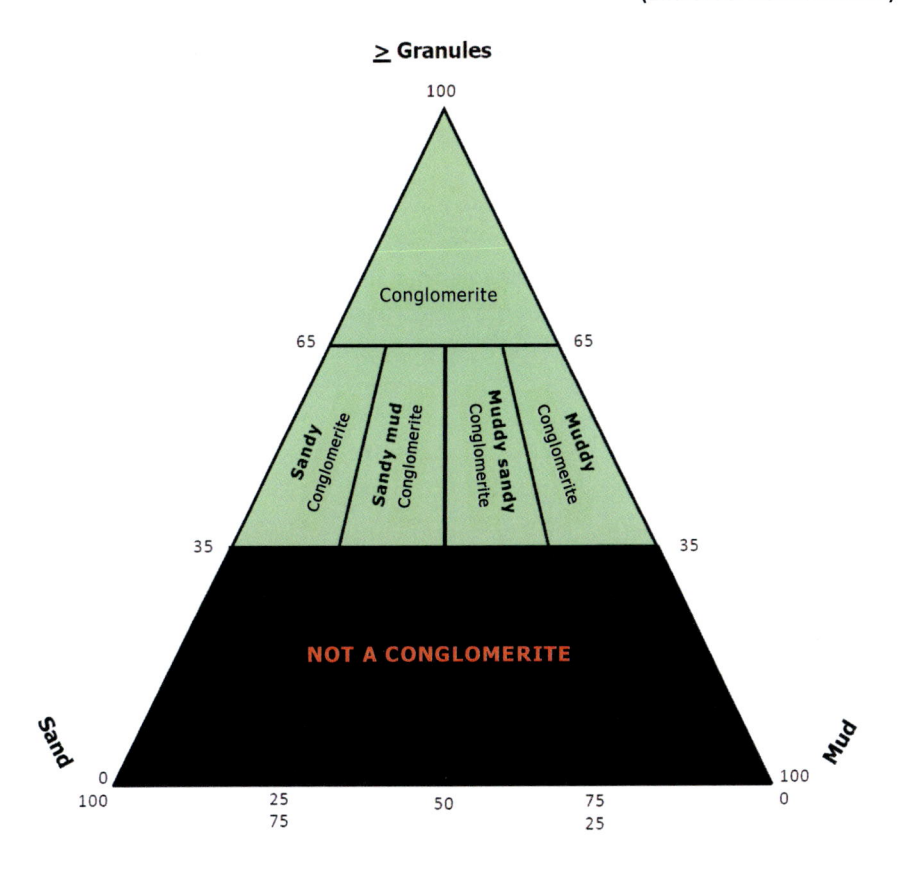

≥ Granules
100

Conglomerite

65 | 65

Sandy Conglomerite | Sandy mud Conglomerite | Muddy sandy Conglomerite | Muddy Conglomerite

35 | 35

NOT A CONGLOMERITE

Sand
0
100 | 25 / 75 | 50 | 75 / 25 | 100
0
Mud

NOTES: *All numbers are in percentages. The word "conglomerate" would be substituted for the appropriate type, e.g. paraconglomerite, orthoconglomerite, or dioconglomerite.*

Adapted from: Baumann and Michaels (2011), and Folk (1954)

IRONITE (Metamorphic Banded Iron Formation)

In an attempt to leave interpretation out of the root name of metamorphosed banded iron formations (BIF), mineral assemblages will be used with newly derived terms (see next page) to classify them.

Ironite is a term named herein to BIF rocks that have undergone significant metamorphism, and the term "BIF" is dropped as part of the root name (p. 146-148). What does significant mean? BIF that has only undergone diagenesis and low grade metamorphism need not beholden to the classification chart on the next page. The reason being is that many of the metamorphic diagnostic minerals, can also be primary minerals deposited with the rock. In the case of low grade metamorphism, the sedimentary chart can be used with the addition to "META-" before the root name.

Since BIF often has minor clay minerals within it, it follows a distinct pseudo-progression of mineral assemblages as metamorphism increases (see next page).

If any assemblage of minerals on the next page is recognized in significant quantity (>5% of rock volume) then it is to be included in the root name. Select the name with the most dominant mineral assemblage. If more than one mineral assemblage is present in abundance, then hyphens can be used. The root name is always from most to least abundant *EXCEPT* the root name "IRONITE" last (see next page). e.g. a meta-BIF with 40% hematite, 20% cummingtonite, 15% almandine, and 25% orthopyroxene would be a "PYROXENITOC-AMPHIBOLITIC-GARNITIC IRONITE".

Stability Fields of Minerals in BIF

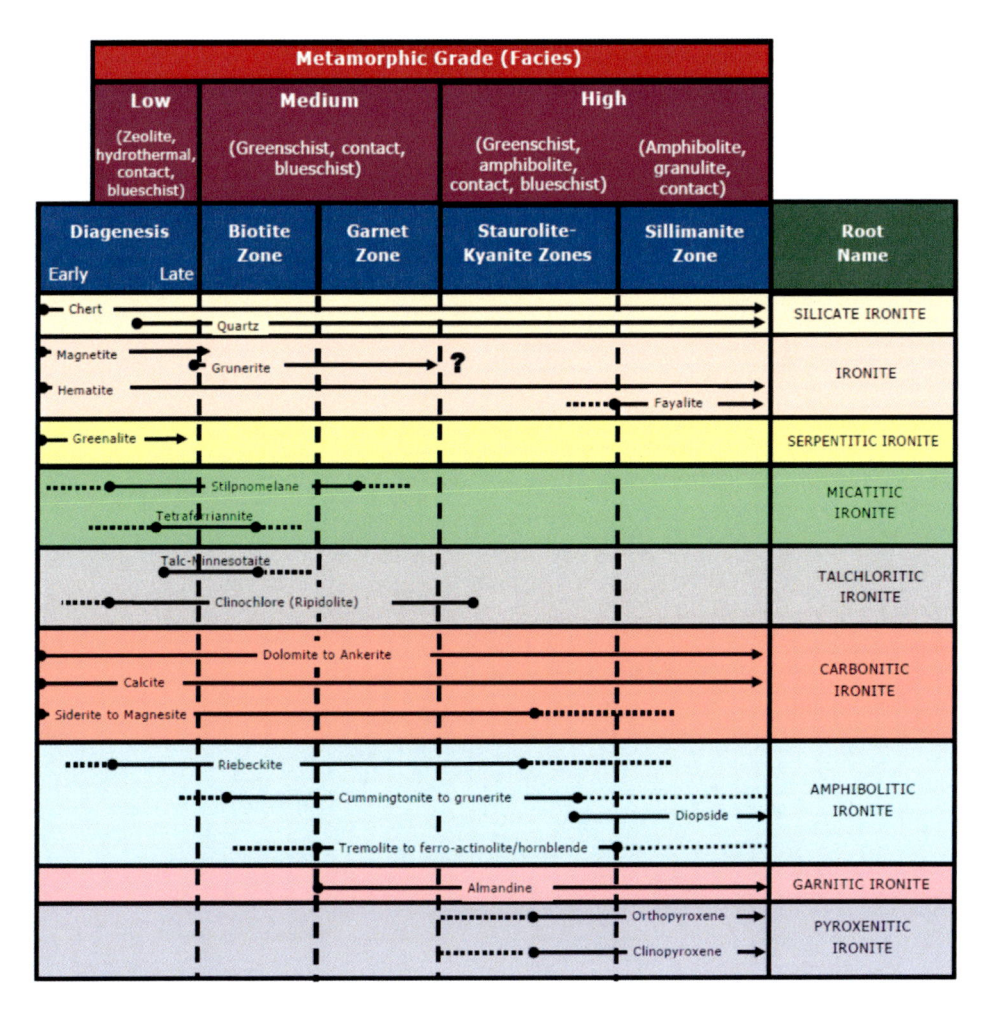

Metamorphic Grade (Facies)					Root Name
Low (Zeolite, hydrothermal, contact, blueschist)	**Medium** (Greenschist, contact, blueschist)		**High** (Greenschist, amphibolite, contact, blueschist)	(Amphibolite, granulite, contact)	
Diagenesis — Early / Late	Biotite Zone	Garnet Zone	Staurolite-Kyanite Zones	Sillimanite Zone	
— Chert — / Quartz —————————————————→					SILICATE IRONITE
Magnetite —→ ; Grunerite ——————→ ? ; Hematite ——————————————→ ; Fayalite —→					IRONITE
Greenalite —→					SERPENTITIC IRONITE
Stilpnomelane ——→ ; Tetraferriannite —————					MICATITIC IRONITE
Talc-Minnesotaite ——— ; Clinochlore (Ripidolite) ——→					TALCHLORITIC IRONITE
Dolomite to Ankerite ————————————————→ ; Calcite ————————— ; Siderite to Magnesite —————————————					CARBONITIC IRONITE
Riebeckite ——— ; Cummingtonite to grunerite ———————— ; Diopside —→ ; Tremolite to ferro-actinolite/hornblende ———					AMPHIBOLITIC IRONITE
Almandine —————————					GARNITIC IRONITE
Orthopyroxene —→ ; Clinopyroxene —→					PYROXENITIC IRONITE

Adapted from: Foley (2009)

<u>NOTE:</u> This chart is only to be used for medium to high grade metamorphosed BIF. Low grade is included for reference only. For low grade metamorphic BIF, see the next page.

The stability of grunerite isn't known past the medium grade of metamorphism.

Ironite Ternary Plot

The below Ironite chart is to be used for metamorphosed banded iron formations (BIFs). It is based off the relative amounts of iron, silica, and carbonate in the rock in the Sedimentary book (p. 223-226)

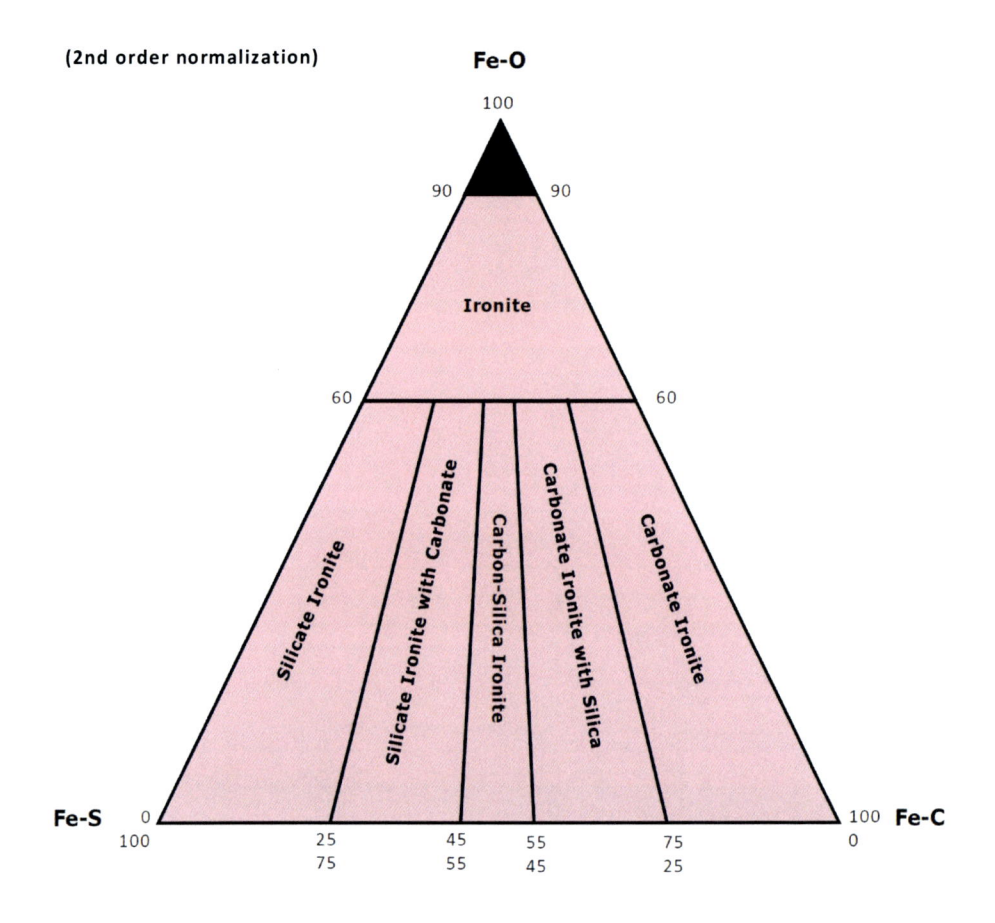

(2nd order normalization)

Fe-O

Ironite

Silicate Ironite
Silicate Ironite with Carbonate
Carbon-Silica Ironite
Carbonate Ironite with Silica
Carbonate Ironite

Fe-S
Fe-C

The above chart is presented herein.

Marble ternary plot based off mineralogy

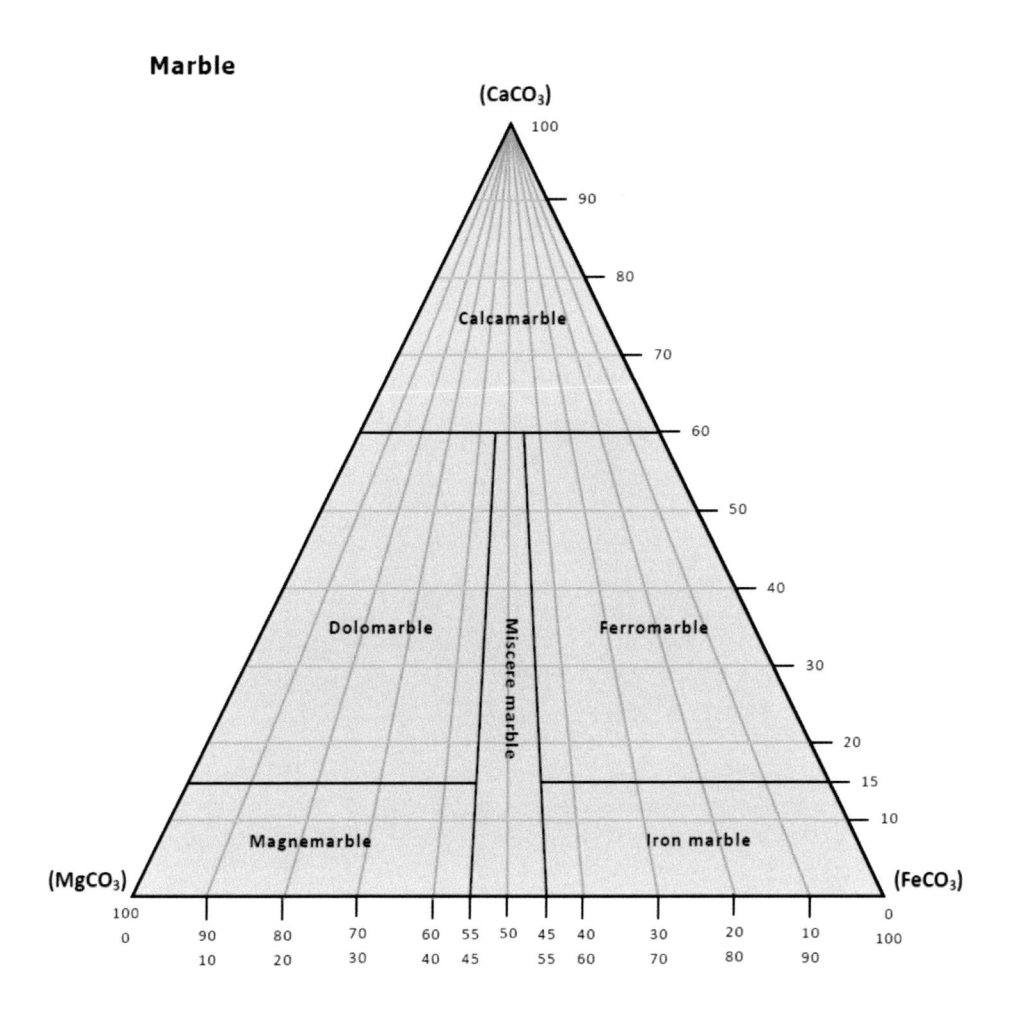

(2nd order normalization)

Miscere is Latin for mix

PART IV:
UNIQUE METAMORPHIC ROCKS

MIGMATITE

Migmatite means "mixed rock". It is a macroscopic composite silicate heterogeneous rock. It commonly has a mafic gneissic component (melanosome) and a lighter felsic portion (leucosome) (Pawley, et al., 2013). Migmatite can appear layered or the leucosomes may occur as pods or form a network of cross cutting dikes, veins, or sills (adapted from Winter, 2001).

Migmatite forms under intense heat and pressure to where some, but not all the rock has melted and recrystallized. There has been some argument to make migmatite a fourth rock type (igneous, sedimentary, metamorphic, and migmatite), but in my opinion, it's an unnecessary complication. They are a combination of metamorphic and igneous rock. Migmatites are more of a "gray area" than a fourth major rock type. There is no need for a new rock type to describe them as they are derived from the highest possible metamorphism without melting the entire rock.

First, a clast or series of clasts brought up by a pluton that was derived from the country rock at depth, is not a migmatite. It's a clastic igneous rock. The clasts in such igneous rocks will often have rims and/or baked zones around the edges of them. Some granites can appear to have magmatic parts as well. Sometimes as a granite cools and culmination, differentiation, or convection occurs on the magma chamber, minerals that have already cooled can appear to be migmatite, but they are not. A migmatite must have been clearly derived from partial melting of a metamorphic rock and not formed from a newly cooled magma body.

Types of migmatite (arterite and venite migmatite)

In order to keep the classification of migmatite simple, only two types (root names, p. 133) are proposed herein. They are "**arterite migmatite**" and "**venite migmatite**", which were first proposed by Wimmenauer and Bryhni (2007).

Using the applicable textural modifiers on p.133 is acceptable.

Arterite Migmatite

Arterite migmatite is when the melanosome (dark mafic parts) parent metamorphic rock is injected by veins, dikes, or sills of leucosome (lighter felsic parts) of igneous origin. These are the most common migmatite in the Archean magmatic terranes of the Great Lakes Region in Ontario, Canada.

Rock is an ARTERITE MIGMATITE on Ontario Route 101, west of the junction with Route 129.

Venite Migmatite

Venite migmatite is when leucosome is derived directly from the melting of the melanosome parent metamorphic rock. Felsic rocks melt before mafic ones, according to Bowen's Reaction Series. That's why leucosome is the first melted. This often can be hard to pick out. There rocks look like highly deformed gneisses and are usually derived from them. Venite migmatite is present in northern Wisconsin and the Upper Peninsula of Michigan, and less common in Ontario (but they are present).

Photo of a VENITE MIGMATITE taken on Ontario Route 101 near the junction with Route 129, near the Sudbury-Algoma District line.

SKARN

Skarn is a coarse grained and highly variable usually carbonate based or iron rich rock. The term derived from Swedish miners to describe waste ore in Paleoproterozoic iron ore bearing sulfide deposits in mines. Its meaning has changed over the years, but it is a useful term. Skarns are formed by a process called metasomatism, which is the hydrothermal alteration of the host rock by dissolving some of it and replacing it with one or more new minerals of differing composition. It alters the remaining host rock through both hydrothermal and/or contact metamorphism. Skarn can also be formed by the alteration of carbonate rocks and BIF due to the introduction of non-magmatic fluids of a host rock during contact metamorphism.

Their mineralogy is related to the protolith, usually a carbonate rock.

They can form on the outside of an intrusion (exoskarn) or on the inside of an intrusion (endoskarn). Exoskarn is the most common.

Skarn can often bear precious metals such as gold, silver, and copper. As well as serve as sources for iron, lead, tungsten, and rare earth elements.

Skarn types/root names and the minerals they include are:

Ca skarn (wollastonite, diopside, grossular, zoisite, anorthite, scapolite, margarite)

Ca/Fe skarn (hedenbergite, andradite, ilvaite, copper, gold, silver)

Cu Skarn (molybdenum, wollastonite, gold, silver)

Mg skarn (forsterite, humites, spinel, phlogopite, clintonite, fassaite)

Mn skarn (rhodonite, tephroite, piemontite)

Ca = calcium, Cu = copper, Fe = iron, Mg = magnesium, Mn = manganese

Adapted from: Bucher and Grapes (2011), Soloviev and Kryazhev (2017)

There are at least three known settings for skarns:

1) Ocean arc settings which tend to bear Ca, and Ca/Fe skarns in gabbro and syenite. The basalts of the Midcontinent Rift also hosts Fe skarn.
2) Continental margin for Mg skarns in granitic and carbonate rocks.
3) Plutons in carbonate rich older continental margins for Cu and Mn skarns in granodiorite and quartz diorite. The basalts of the Midcontinent Rift also hosts Cu skarn.

Adapted from: Soloviev and Kryazhev (2017)

Using the textural and root modifiers on p.133 is acceptable where applicable.

Native copper and white quartz in an "Fe SKARN" from the Upper Peninsula, Michigan. The protolith is a basalt.

Metamorphic rocks under high magnification

This is what is commonly called a greenstone. It's from Ontario, Canada. It is actually an ultramafic rock. Likely a meta pyroxenite.

2cm

Diameter ~3.5 mm

Although this rock is green in color, olivine is only a minor component of this rock. Most of the dark green color is epidote and pyroxene. The lighter green is chlorite due to metamorphism. Most of the white in the upper left is plagioclase.

Diameter ~1 mm

The darkest color is mostly pyroxene. The medium greens are olivine and epidote. The light greens and yellowish colors are chlorite. Here the white plagioclase stands out a bit more than in the picture above.

Metamorphic rocks under high magnification

This is a gray iron carbonate that has a red agate seam. The rock is slightly metamorphosed. It is the Gunflint Formation from near Thunder Bay Ontario.

Diameter ~4 mm

Here is a close up of the red microcrystalline quartz and translucent white quartz that makes up the agate part of the rock.

Diameter ~1 mm

Here you can see the alternating white and red colors of the agate.

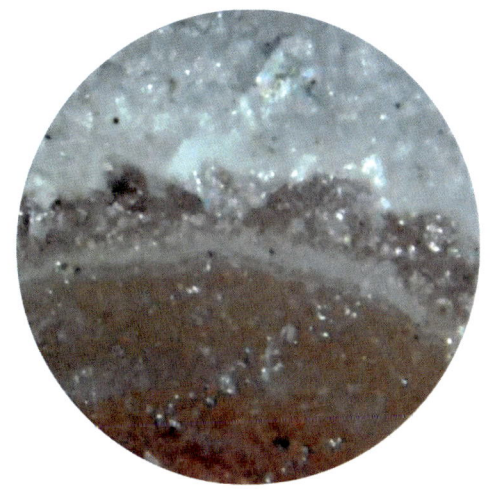

Metamorphic rocks under high magnification

This is a ferrous proto-quartzite from Marquette County in the Upper Peninsula of Michigan. The diameter of the circle is about 1mm. In it you can see white translucent quartz. The reds are actually mostly feldspar. The nearly black is a combination of feldspar and quartz rich in iron.

This is rare type of proto-quartzite from Elliot Lake Ontario called the Mississagi Formation. The diameter of the circle is about 3.5mm. This formation is pre-oxygen in the atmosphere. It formed in a reducing environment. The yellow is mud and the light translucent gray is a mixture of quartz and feldspar. Most of the black is iron minerals, but sone are a radioactive mineral called uraninite (pitchblende).

Metamorphic rocks under high magnification

This is a biotite schist is a foliated rock from Watersmeet Michigan in the Upper Peninsula.

2cm

Diameter ~3.5mm

You can see the well developed foliation here from the upper left to lower right. The white is a mixture of quartz and feldspar. The black is mostly biotite with some pyroxenes. The green was too metamorphosed to be identified without chemical analysis but is likely chlorite or epidote.

Diameter ~1mm

The yellow arrow is pointing to a crystal of orthopyroxene as are the rest of the euhedral crystals. The subhedral black crystals are mostly biotite. Most of the non black minerals are quartz and feldspar.

This page is intentionally left blank

Lab Reference Book
Classifying Sedimentary Rocks

Steven D.J. Baumann, PG

All photos in this book were taken by the author. Cover photos are as follows: Upper left is the carbonate breccia of the upper Randville Dolostone. Upper right is the Jacobsville Sandstone. Lower left is banded iron formation (BIF) from the Upper Peninsula of Michigan). Lower right is a conglomerate from and interflow conglomerate.

INTRODUCTION

This book goes trough the classification of sedimentary rocks in a laboratory setting. I assume you have a basic understanding of rock classification. This book isn't designed to teach you. It is designed to be a reference.

This book also focuses on visual classification techniques, similar to what is outlined in the "Lab Reference Book; Classifying Igneous Rocks (Visual Methods)", Baumann (2019). This book is also not an engineering soil lab book. Things like sieve analysis, hydrometer tests, Atterberg limits, etc., will not be included herein.

This book is structured a little bit differently than the igneous lab book. It is divided into four sections. They are "general information", "clastic rocks", "non clastic rocks" and "references and notes". The "general" section deals with things that apply across the board such as fabric, bedding, and grain size. "Clastic rocks" deals with your sandstones, conglomerates, and mudstones. The third part is a bit more complicated. The "non clastic rocks" deal with both chemical and biological / organic rocks. It includes such things as, carbonate, coal, and evaporites. The fourth section is "references and notes", here you can add your own notes and see what sedimentary rocks look like under high magnification.

Some of the classification schemes herein may be different than ones in my 2019 book "Midwest Field Geologist's Reference Book (Deluxe Edition)". The reason being, is that in a lab setting, descriptions are more detailed as there is actual time to describe the rock in a controlled setting. The field book still stands for use in the field. This book also focuses on the sedimentary rocks of the Midwest.

All mineral data was adapted from the "Handbook of Minerology. Individual mineral data sheets are available for free at: **handbookofmineralogy.org**.

All photos were taken by the author. All drawings were done by the author. I hope you find this book useful.

PART I
GENERAL SEDIMENTARY ROCK TEXTURES

Banded Iron Formation (BIF) from near Alberta, Michigan in the Upper Peninsula.

SEDIMENTARY ROCK FABRIC

Rock fabric (in a sedimentary context) is the part of a rock's texture in which the characteristics of a rock are controlled by porosity and permeability based on the mutual arrangement of grains. In igneous rocks and in some metamorphic rocks, this is a non issue, because they do not usually transmit fluids like groundwater, unless fractured. In sedimentary rocks, porosity is important. Many things can effect porosity and permeability and thus a rock's ability to move fluids due to gravity. Density, sorting, cementation, bedding, faults, fractures, grain angularity, grain size, lithification, and diagenesis. Most of these things will be touched upon later. This section will just deal with how the grains relate to one another.

The things that will directly effect porosity and permeability are grain sorting, packing, contacts, orientation (or alignment), and support in relationship to the matrix. The **matrix** in this book is defined as "the usually microscopic non grain binding, solid matter in between the discernable grains". This excludes gases (air) and liquids (oil and water) that may occur between grains. Matrix is deposited with the rock as sediment before lithification. Microscopic grains are made of "fines". **Fines** are defined as grains <0.62mm (or passing the #230 U.S. sieve) in size. Often fines are referred to as "mud".

This is different from cement, although the two have been used interchangeably by people. **Cement** is "the micro to macroscopic solid matter in between grains that physically binds grains together". Cement is usually not deposited with the sediment, although it can be. It is usually introduced during lithification or diagenesis by fluids and pressure. However, it can be partially derived from the grains themselves. In sedimentary rocks the most common cements are calcite, silica, and iron oxides (like hematite and limonite).

Grains are perhaps the most important part of a rock's texture. Pages 164-177 deal with sedimentary grains. See p.216-217 for crystalline grains pertaining to carbonate and some chemical rocks.

Grain Sorting

Grain sorting is how closely the grains are in grain size or if they are skewed in favor of one, or more, grain sizes. This does have a quantitative value that will only be touched on here, only the qualitative aspects will be shown. NOTE: *Sorting is the opposite of grading in engineering*. In geology a well sorted sandstone is a poorly graded sandstone in engineering.

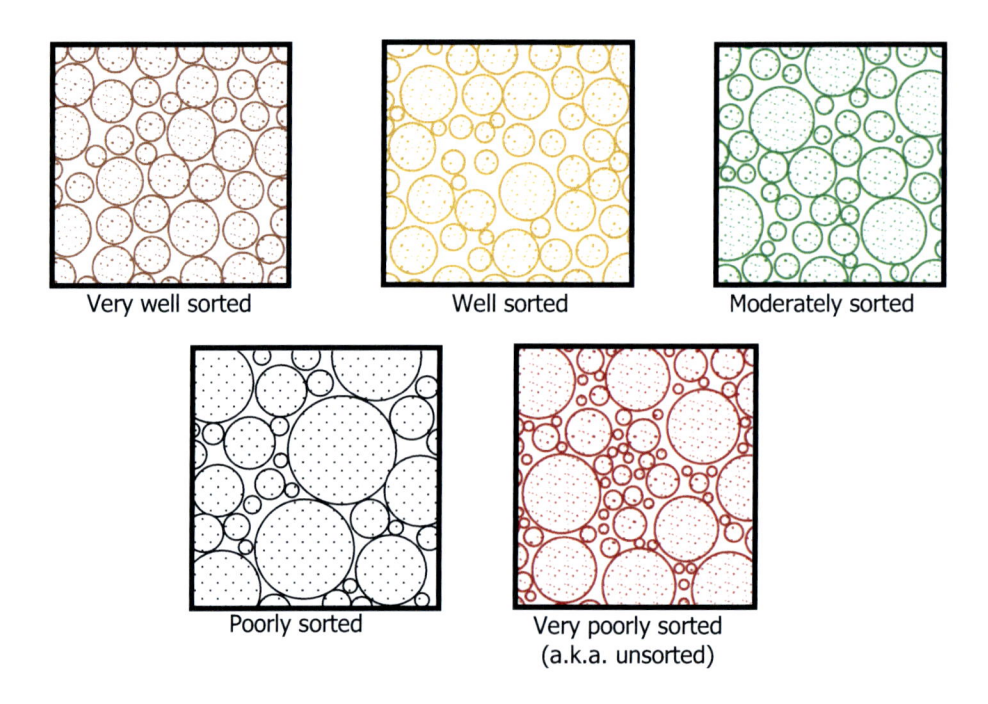

Very well sorted   Well sorted   Moderately sorted

Poorly sorted   Very poorly sorted
(a.k.a. unsorted)

Adapted from: Tucker (2003)

The above diagrams are idealized sorting based of spheres. In reality, that doesn't happen. Below are more realistic depictions of the extremes.

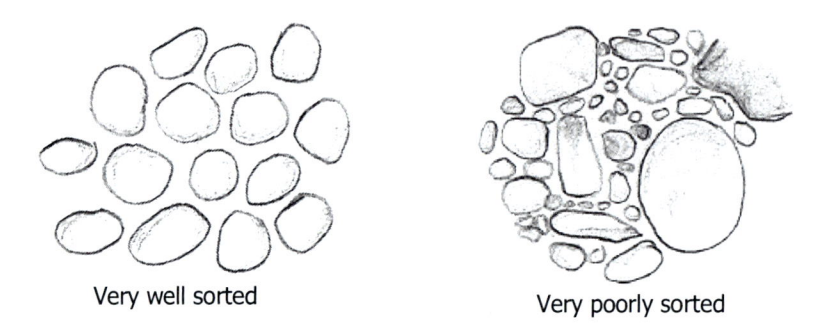

Very well sorted   Very poorly sorted

Intergrain Relationships

Grain Packing

Packing just means the grains are arranged in such away to allow a certain percentage of pore space between them. The below two drawings assume rounded grains. In reality, grains are never this round. Things like shale have very platy grains and very large porosity. Except shale particles are so small, water does not effectively move through them. So although porosity in shale is high, permeability is low.

Cubic packing
(45-50% porosity)

Rhombohedral packing
(25-30% porosity)

Adapted from: Tucker (2003)

Grain Contacts

Contact is just how grains can be in direct contact with one another. In order for this to occur you have to have a grain supported rock. A matrix supported rock won't express this as the grains are not in direct contact with one another.

Point contact

Sutured contact

Overgrowth contact

Nearly all grains touch one another at least one point of contact.

Grains themselves have undergone alteration to the point where they have fused together during diagenesis and low grade metamorphism.

Space in between the grains as fused to the grains during lithification and sometimes diagenesis.

Adapted from: Folk (1980)

Grain Orientation / Alignment

Orientation or alignment (either or is acceptable, just be consistent) is a qualitative description indicating whether or not the grains show a preferable orientation, like in a river. As opposed to a random one like in a debris flow.

Slight Orientation or Multidirectional Alignment

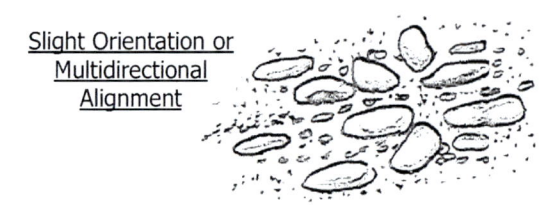

Random Orientation or Jumbled Alignment

Preferred Orientation or Directional Alignment

Grain Support

This is just a relative visual arrangement of the grains in relationship to the matrix. Remember, the matrix is "the microscopic solid matter in between the discernable grains". This excludes gases (air) and liquids (oil and water) that may occur between grains.

Grain supported

When >1/2 to almost all the grains are in physical contact with one another.

Matrix supported

When 1/10 to 1/2 of the grains are in physical contact with one another.

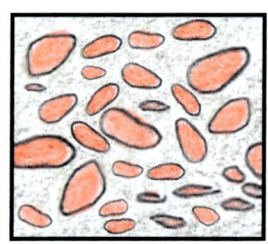

Suspended

When <1/10 of the grains are touching but the grains aren't so far apart that the matrix dominates.

Grains

This part deals with the physical characteristics of grains, e.g. shape, size, percentage. Understanding the physical characteristics of grains can aid in determining origin. Although not conclusive without the accompanying field observations, some generalizations about grains can be made.

1) Generally, the smaller and more rounded a grain is, the further total distance it has traveled. What does this mean? It's total distance. Not straight distance from origin to observed resting place. A grain of sand could be on a beach a couple of meters from the outcrop it was derived from. Yet, the sand grain could have traveled many kilometers as it was repeatedly smashed up on the rock, then down on the beach, then back on the rock, because of the waves and tides. Or it can travel several kilometers in a single storm event and still be large and angular.

2) The more mature (quartz rich) a deposit is, the more the grains have been recycled. Things like lithic arenites tend to be derived locally and only transported once. As where quartz arenites could be in a 2nd, 3rd, etc., generation cycle.

3) Organic deposits like pure coal and bio-carbonates are in situ deposits. They are not generally recycled deposits like sandstones. Organic rocks, like coal, are not classified based on grain texture.

4) The shape of a grain can also indicate an origin (excluding clay). The flatter and longer the grain, the more likely it was deposited in stable static conditions.

5) Mechanical weathering can only break down a rock so far before it becomes immune to further weathering because of conditions on Earth's surface. That smallest particle size that can be mechanically broken down is silt. The smaller the grains in a mechanically weathered rock, the more likely the rock is to be pure quartz. Clay particles form from chemical alteration, not from mechanical weathering. So the largest clay particles can be bigger than the smallest silt particles. So there is an overlap from about 0.002-0.004mm.

6) Smaller grains tend to be more rounded. There are certain environments where this doesn't hold, like alluvial fans.

Grain Angularity

The angularity of a grain is very important to the descriptive process. It is *NOT* necessary to give an angularity for each grain. Ranges for the overall rock can be given instead of detailed break downs.

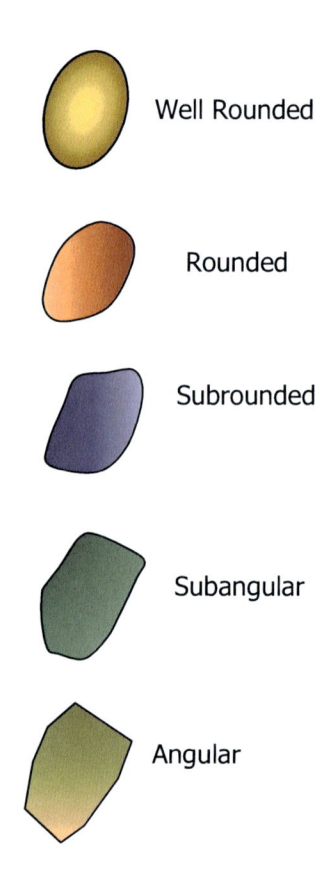

Well Rounded

Rounded

Subrounded

Subangular

Angular

A clastic rock is one that has been derived by mechanical weathering of other rocks. The most common clastic grain shapes are rounded to subangular.

If you want, relative percentages of grain angularity based on volume, can be broken out. For example, if you have a quartz arenite with no matrix you can say, "10% rounded fine grains, 50% subrounded medium grains, and 40% subangular coarse grains". Or you can just say "subangular to rounded".

Grain Size

This is perhaps the defining characteristic in sedimentary rock description. You can give detailed percentages, but like with angularity, a range can be given. Geologists use the Wentworth Scale. The chart on page 11 includes the Unified Soil Classification System (USCS) for comparison purposes.

Magnification

Most geologists will use a 10x or 15x magnification hand lens (a.k.a. jeweler's lens / tiny magnifying glass) in the field. For lab identification you need more powerful magnification.

What do things like 10x mean? Well, it means the image is magnified 10x the original size, but it's not a linear 10x. A 1mm long object will not appear to be 10mm in a 10x hand lens. It will appear to be 3.16mm long. That's because you are working with surface area, NOT linear length. You are working with 2 dimensions not 1. With a 40x lens the smallest particle visible to the naked eye (0.0625mm) is going to appear to be 0.3953mm long.

How is this stuff calculated? You need the square root of your magnification. Depending on if you want to calculate, you will either multiply or divide by the desired particle size.

If you want to know how large a particle size will look under magnification, you use the following equation:

X = magnification   Φ = comparative particle size $\sqrt{X} \,(\Phi) = Y$
    Y = desired particle size

Example: You want to know how long a 0.5mm grain will look under a 20x lens

$\sqrt{20} \,(0.5mm) = 2.24mm$

If you want to know the smallest resolution that you can see under magnification, you use the following equation:

X = magnification   0.0625mm = smallest visible particle size $Y = (0.0625mm) / \sqrt{X}$
    Y = desired particle size

Example: You want to know the smallest particle visible under 20x lens.

$0.0114mm = (0.0625mm) / \sqrt{20}$

10x = 0.01976mm	40x = 0.009882mm
20x = 0.01398mm	60x = 0.008069mm
30x = 0.01141mm	120x = 0.005705mm

My 40x hand lens for lab use. It has LED lights on the bottom.

Angular paraconglomerate northwest of Calumet, Michigan in the Upper Peninsula under the 40x hand lens in the above photo.

172

Wentworth Scale

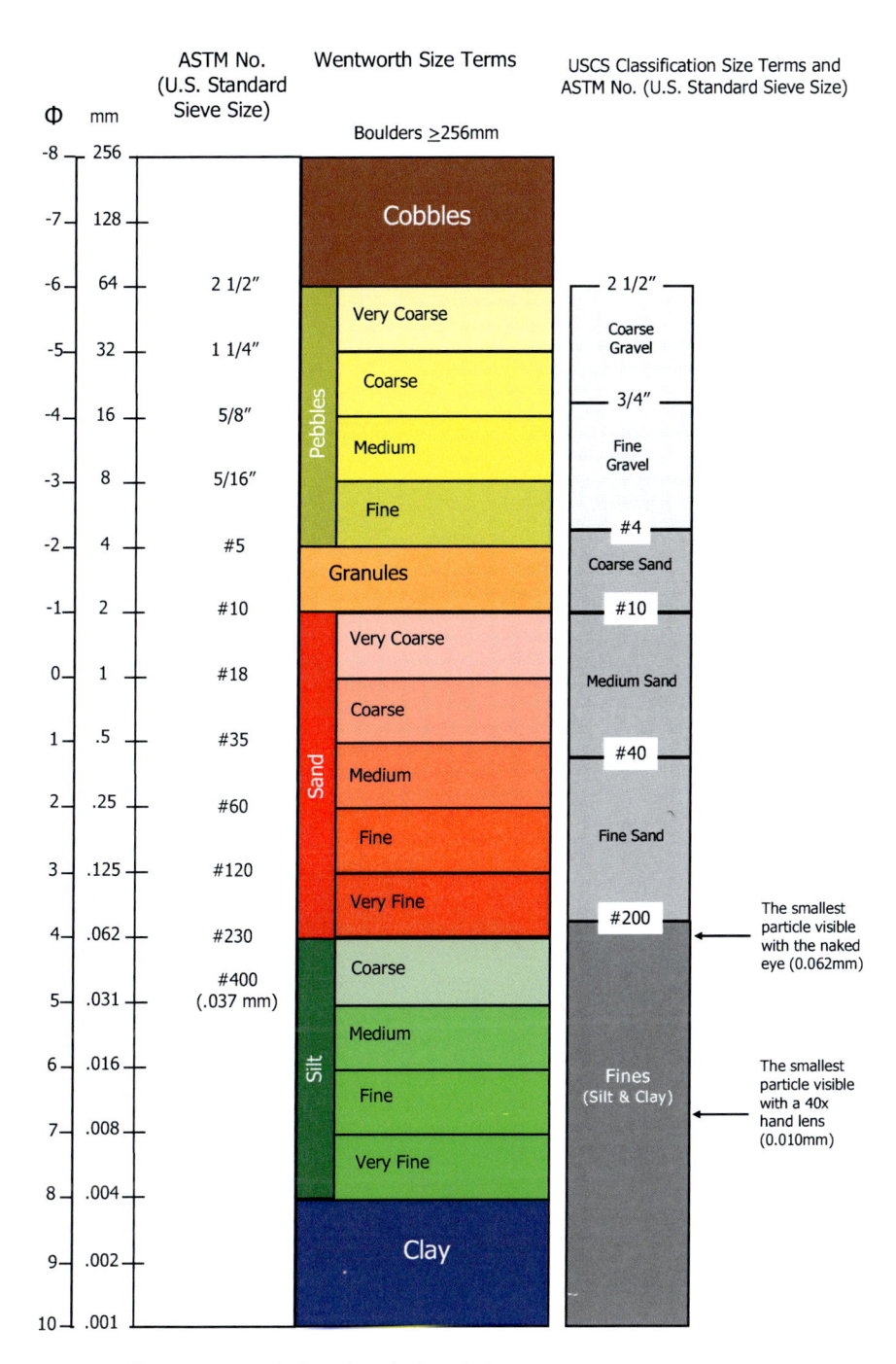

Φ	mm	ASTM No. (U.S. Standard Sieve Size)	Wentworth Size Terms	USCS Classification Size Terms and ASTM No. (U.S. Standard Sieve Size)
-8	256		Boulders ≥256mm	
-7	128		Cobbles	
-6	64	2 1/2"	Pebbles — Very Coarse	2 1/2" — Coarse Gravel
-5	32	1 1/4"	Coarse	
-4	16	5/8"	Medium	3/4" — Fine Gravel
-3	8	5/16"	Fine	
-2	4	#5	Granules	#4 — Coarse Sand
-1	2	#10	Sand — Very Coarse	#10 — Medium Sand
0	1	#18	Coarse	
1	.5	#35	Medium	#40 — Fine Sand
2	.25	#60	Fine	
3	.125	#120	Very Fine	
4	.062	#230	Silt — Coarse	#200 — Fines (Silt & Clay)
5	.031	#400 (.037 mm)	Medium	
6	.016		Fine	
7	.008		Very Fine	
8	.004		Clay	
9	.002			
10	.001			

The smallest particle visible with the naked eye (0.062mm)

The smallest particle visible with a 40x hand lens (0.010mm)

Adapted from: Wentworth (1922) and ASTM 2487

Particle size comparisons

These are to scale.

Granule / Very Coarse Sand

0.07874" = 2mm

Very Coarse Sand / Coarse Sand

0.03937" = 1mm

Coarse Sand / Medium Sand

0.01969" = 0.5mm

Medium Sand / Fine Sand

0.00943" = 0.25mm

Fine Sand / Very Fine Sand

0.004921" = 0.125mm

Very Fine Sand / Coarse Silt

0.002441" = 0.062mm

Grain Shape

The shape of a grain is important for understanding origin of the particles. It is usually not even mentioned in most petrographic analysis, although it can be included, if needed. Below are basic grain shapes. Obviously grains in nature aren't this perfect. Just pick the one that is the closest. The below examples do not take into account the bending of a grain along one or more axis.

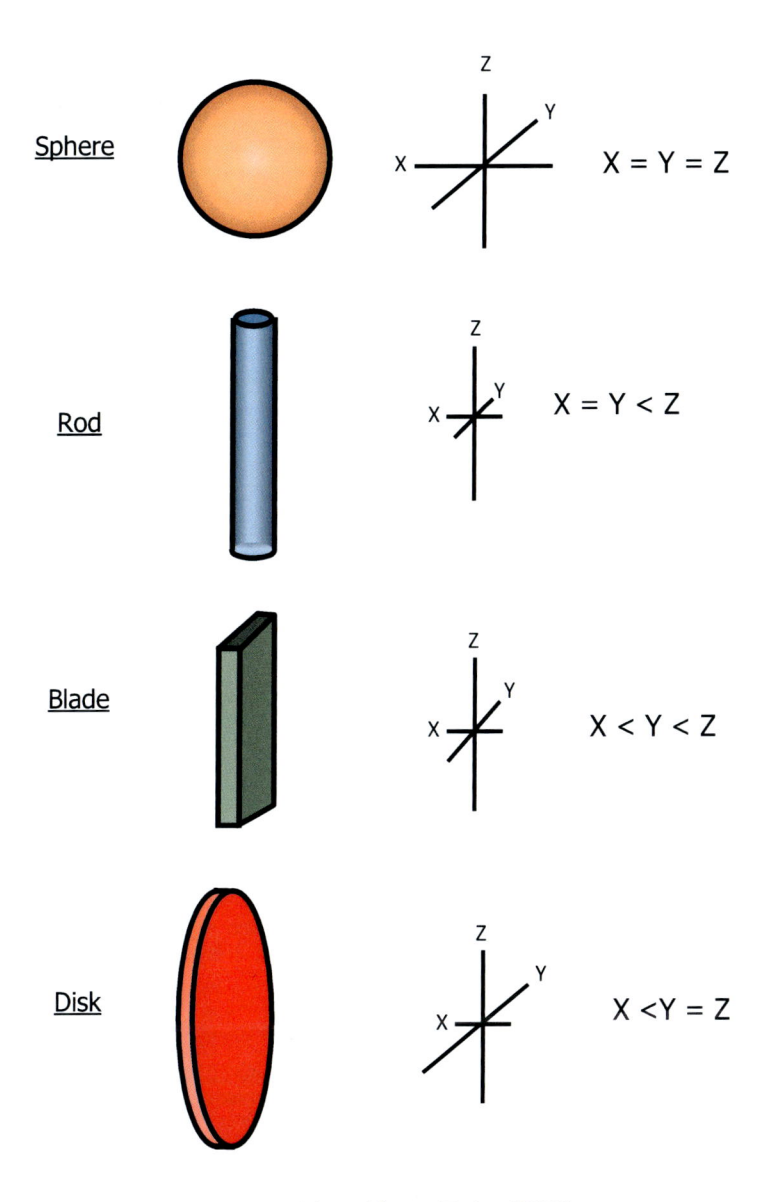

Sphere — $X = Y = Z$

Rod — $X = Y < Z$

Blade — $X < Y < Z$

Disk — $X < Y = Z$

Adapted from: Tucker (2003)

Grain Distribution

Grain distribution is a function of two things; grain size and sorting. Well sorted rocks will be nearly all the same grain size and exhibit normal distribution. As where poorly sorted rocks with varying grain sized will exhibit bimodal or multimodal distributions. Skewness is how much the peak(s) (a.k.a. frequency curve) deviate from center. Kurtosis is how pointy the peak is. When you have computed percentages of grains, you plot them on a histogram of frequency % (y-axis) vs. grain size (x-axis), then you generate a curve. From there you can make interpretations. Curves do *NOT* have to be unimodal. They can be bimodal, trimodal, or even exhibit smaller bumps on the main curve.

Skewness

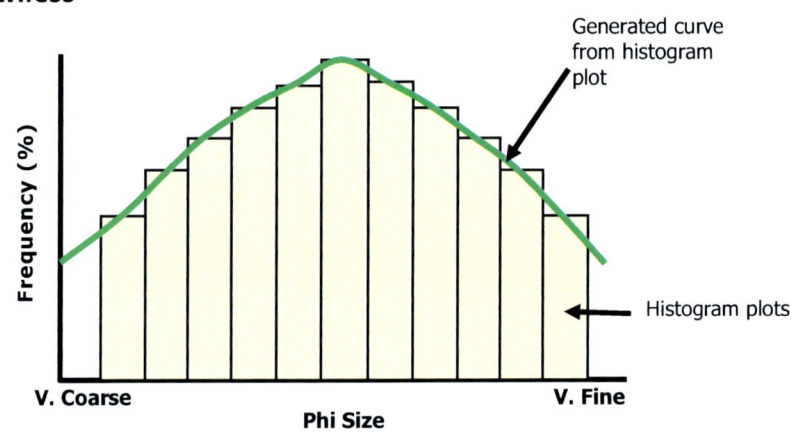

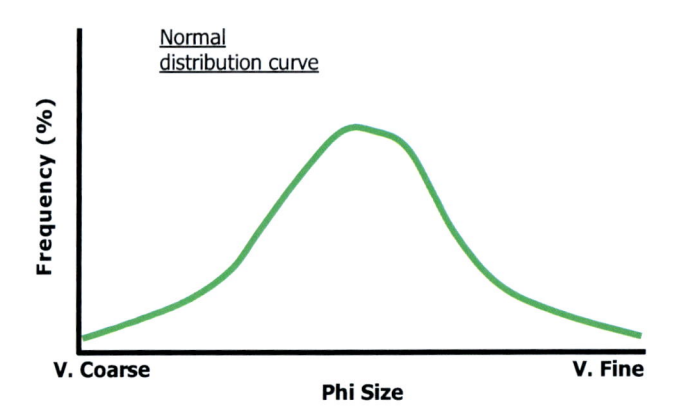

Adapted from: Boggs (2009)

Positive skew indicates coarse grains exhibit better sorting than fine

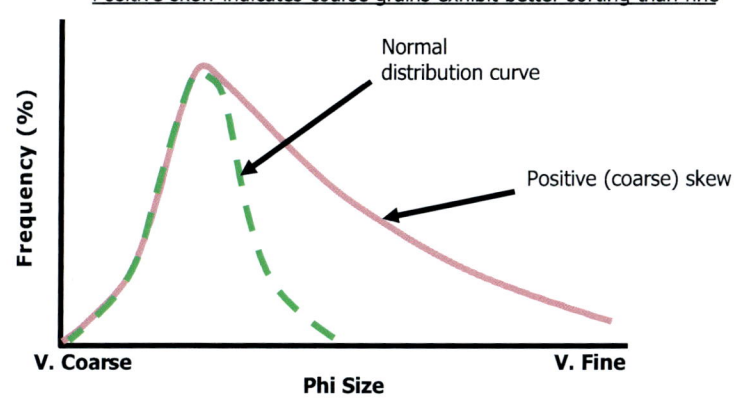

- Normal distribution curve
- Positive (coarse) skew

Frequency (%) vs **Phi Size** (V. Coarse → V. Fine)

Negative skew indicates fine grains exhibit better sorting than coarse

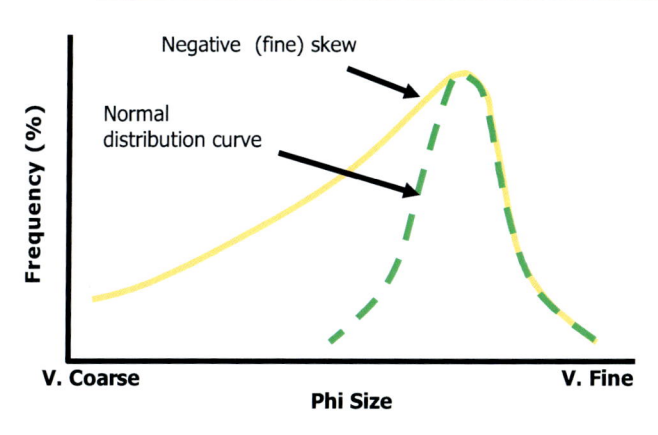

- Negative (fine) skew
- Normal distribution curve

Frequency (%) vs **Phi Size** (V. Coarse → V. Fine)

Kurtosis

- Leptokurtic
- Normal
- Platykurtic

Frequency (%) vs **Phi Size** (V. Coarse → V. Fine)

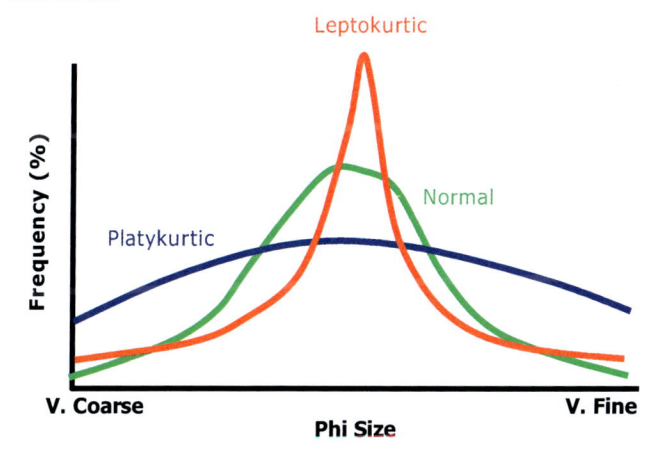

Adapted from: Boggs (2009)

BEDDING

Bedding in the field is complicated as you can have beds or laminations within beds called "bedsets". A rock in a box is a bit different. It is smaller and generally confined to a single bedset, if any. If bedding planes can be identified within a hand sample, they should be as well. The word "lamination" can be used in lieu of the word "bed", if appropriate. Although bedding is usually determined in the field, it can be determined from a hand sample, it should. If not, then it is left out. Silence is golden.

Bedding Planes

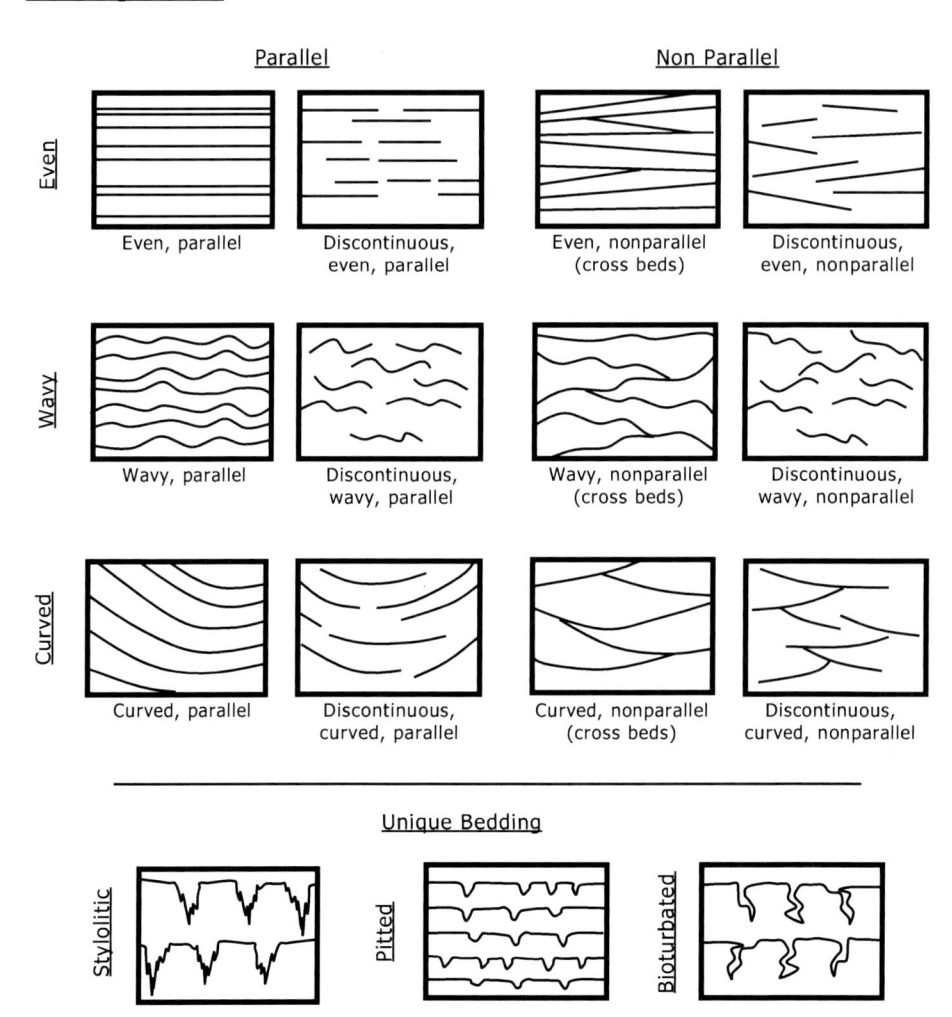

	Parallel		Non Parallel	
Even	Even, parallel	Discontinuous, even, parallel	Even, nonparallel (cross beds)	Discontinuous, even, nonparallel
Wavy	Wavy, parallel	Discontinuous, wavy, parallel	Wavy, nonparallel (cross beds)	Discontinuous, wavy, nonparallel
Curved	Curved, parallel	Discontinuous, curved, parallel	Curved, nonparallel (cross beds)	Discontinuous, curved, nonparallel

Unique Bedding

Stylolitic | Pitted | Bioturbated

Adapted from: Campbell (1967) and Greensmith (1989)

Cross beds

Cross bedding is usually at a macro scale. I feel that they are important enough to be included in this lab reference although they are rarely at a hand specimen scale.

Cross bedding is a type of bedding and a primary structure in sedimentary rocks. I will be describing cross beds, which can be made of cross laminations as well, but I will not be using the term lamination. There is no solid consensus on how to classify cross beds. In my effort to leave out interpretations in the description process, I will be grouping them by structure.

It's important to remember that certain patterns and grain size stacking can lead to the correct environmental interpretations. Cross beds can be in any sedimentary rock. They can also be in meta-sedimentary and sometimes even in igneous rocks. As a result I will not be addressing grain or crystal size. Just the shape in cross section.

Cross beds are different from normal bedding because they truncate or highly deform adjacent beds. They form during fluid flow. That fluid is usually sea waves or rivers. But it can also be aeolian or even igneous melt. All cross beds are defined by their bedding planes, in cross section. However the surface of the bedding planes can also be important. I will not be addressing that here. Although sedimentary petrology dives deep into the subject.

The chart expressing my classification system is on p. 182.

Planar cross beds are at angles to one another with little to no curving. They are usually conducive of an aeolian depositional environment or shallow high energy water environments. They consist of tabular, herring bone, and bounded cross beds.

Tabular cross beds tend to form at relatively steep angles and on a large scale. They are deposited in aeolian environments.

Herring bone cross beds are at alternating angles bound by level bed sets that resemble the ribs of a fish. They form in strong alternating tidal environments, that's why the direction of the herring "ribs" change direction from the bed sets above and below.

Bounded cross beds exist in a similar manner as herring bone cross beds except the angled beds all lean in the same direction. These types of beds usually form in low energy environments with a constant direction of flow. They can be shallow marine or nearshore aeolian.

Tangential cross beds are the second group. The angles are curved, there's no planar beds except at the bet set level. These cross beds are nearly all water lain.

Hummocky cross beds are formed by unidirectional and oscillatory flow are usually formed in shallow marine, large lake bed, or very wide braided stream beds that are influenced by waves during storm events. They sort of look like deep plates stacked on top of one another right side up and upside down. They also form in nearshore to shoreline environments where large storms being sand into tidal environments. They can also form in glacial environments known as "hummocky moraines" (Chandler et al., 2021).

Trough cross bedding is scoured out and scoop shaped then filled with sediment. The filled in beds can follow the scoops or cross bed filled all in one direction, like in flaser bedding. The troughs are usually formed as terrestrial channel fill in braided stream deposits or subaqueous nearshore environments.

Flaser cross bedding is similar to trough cross bedding except the bedsets are bounded by finer (usually mud sized) particles. They form in environments that have alternating high energy and low energy water environments, such as in seasonal fluctuations.

Contrary cross beds express a highly disturbed, yet mostly ordered disturbance, that interfere with neighboring beds. These types of beds are only in deep water environments.

Disk and ribbon cross beds are a semi-organized system of laminations and beds that tumble over one another. And grain size can vary greatly. They are formed when a high energy volume of water hits a stagnant one. Like in a turbidity flow or a subaqueous slump.

Ball and pillow cross beds are rounded balls of sediment between somewhat level beds. The balls appear internally parallel and laminated but are actually elliptical from top view. These are expected to from mainly as load structures when sand is quickly deposited on top of mud. This cah happen in a tectonic or debris flow deep subaqueous setting.

Combination cross bedding consists if a combination of features from both tabular and tangential cross bedding. We can spend all week sdding and dividing to this list, depending on how far down the rabbit hole we want to go. But simplicity is best

Swaley cross bedding is a combination of hummocky and bounded cross beds. They are mostly planar but curve at bed sets and are not angled as steeply as planar or most tangential cross beds. They form in marine or lacustral subaqueous environments just below the wave base.

Sigmoidal cross beds are shaped like lenses bent into a smooth S-shape and bounded by planar and usually level bed sets. They an form in different sub aqueous environments but are typically formed due to the progradation of deltas.

Folded cross beds are sort of a waste bin term for cross beds that exist in nearly planar bed sets but internally express unidirectional preferred flow tha bends over itself forming sideways "U's" with the bottom of the "U" pointing downstream. These can only form in deep water environments where existing cross beds are folded over themselves when by a slide in lateral movement, caused by tectonic activity, before being buried by heavier and more level sediments.

Cross bed patterns

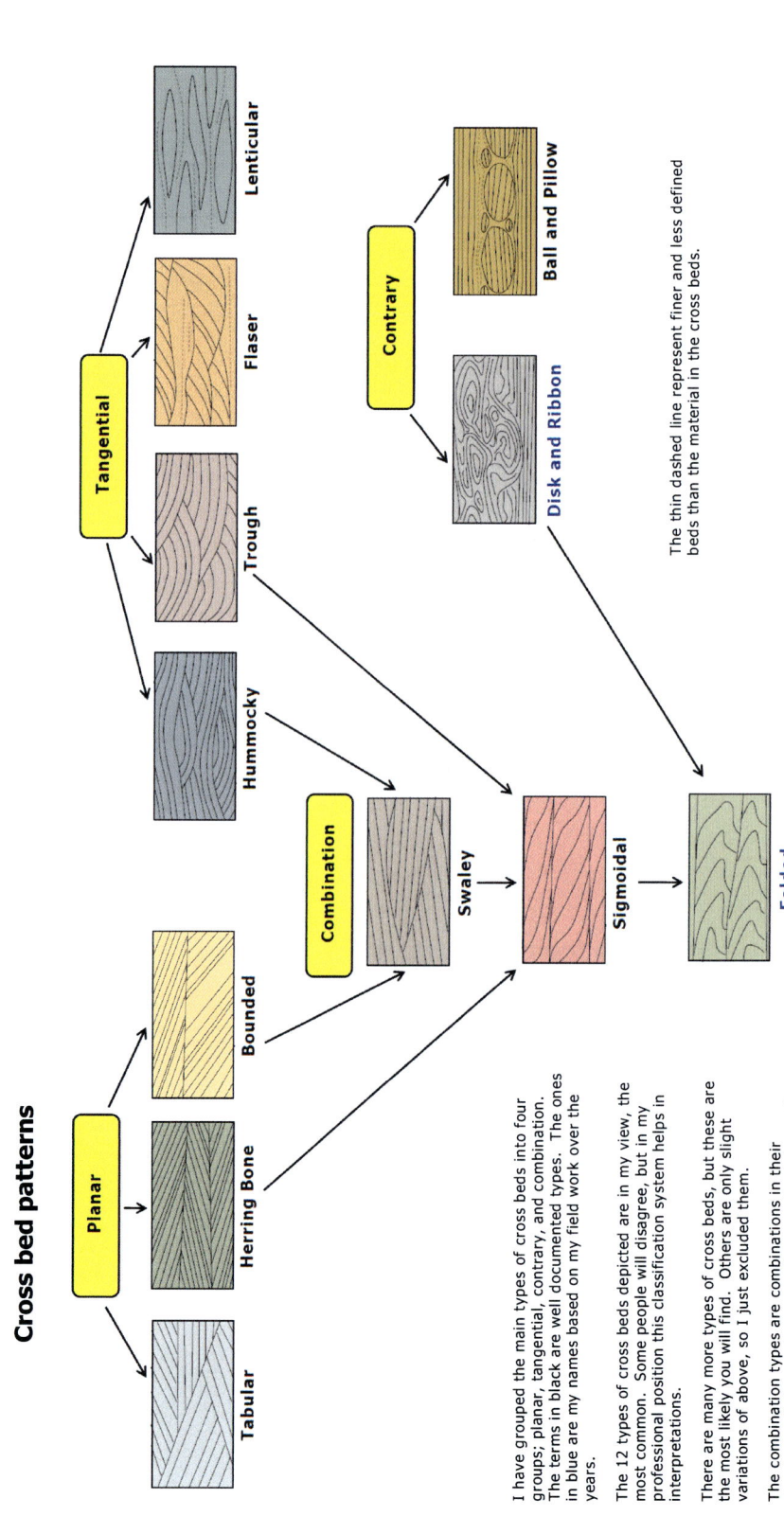

Planar: Tabular, Herring Bone, Bounded

Tangential: Hummocky, Trough, Flaser, Lenticular

Combination: Swaley

Contrary: Disk and Ribbon, Ball and Pillow

Sigmoidal → Folded

The thin dashed line represent finer and less defined beds than the material in the cross beds.

I have grouped the main types of cross beds into four groups; planar, tangential, contrary, and combination. The terms in black are well documented types. The ones in blue are my names based on my field work over the years.

The 12 types of cross beds depicted are in my view, the most common. Some people will disagree, but in my professional position this classification system helps in interpretations.

There are many more types of cross beds, but these are the most likely you will find. Others are only slight variations of above, so I just excluded them.

The combination types are combinations in their structure. It does not necessarily mean the interpretive environments are combined as well.

Adapted from: Tucker, 2006

Bedding Thickness Terms

A lot of different people have different concepts of what a bed and a lamination are. This is what I use, derived and altered from Campbell (1967). Very thick laminations and very thin bedding fall within the same range (see below). Use your best judgment to pick one or the other. If a rock is mostly laminated, you probably want to go with the lamination nomenclature and not the bedding nomenclature, unless they are within greater "bed sets".

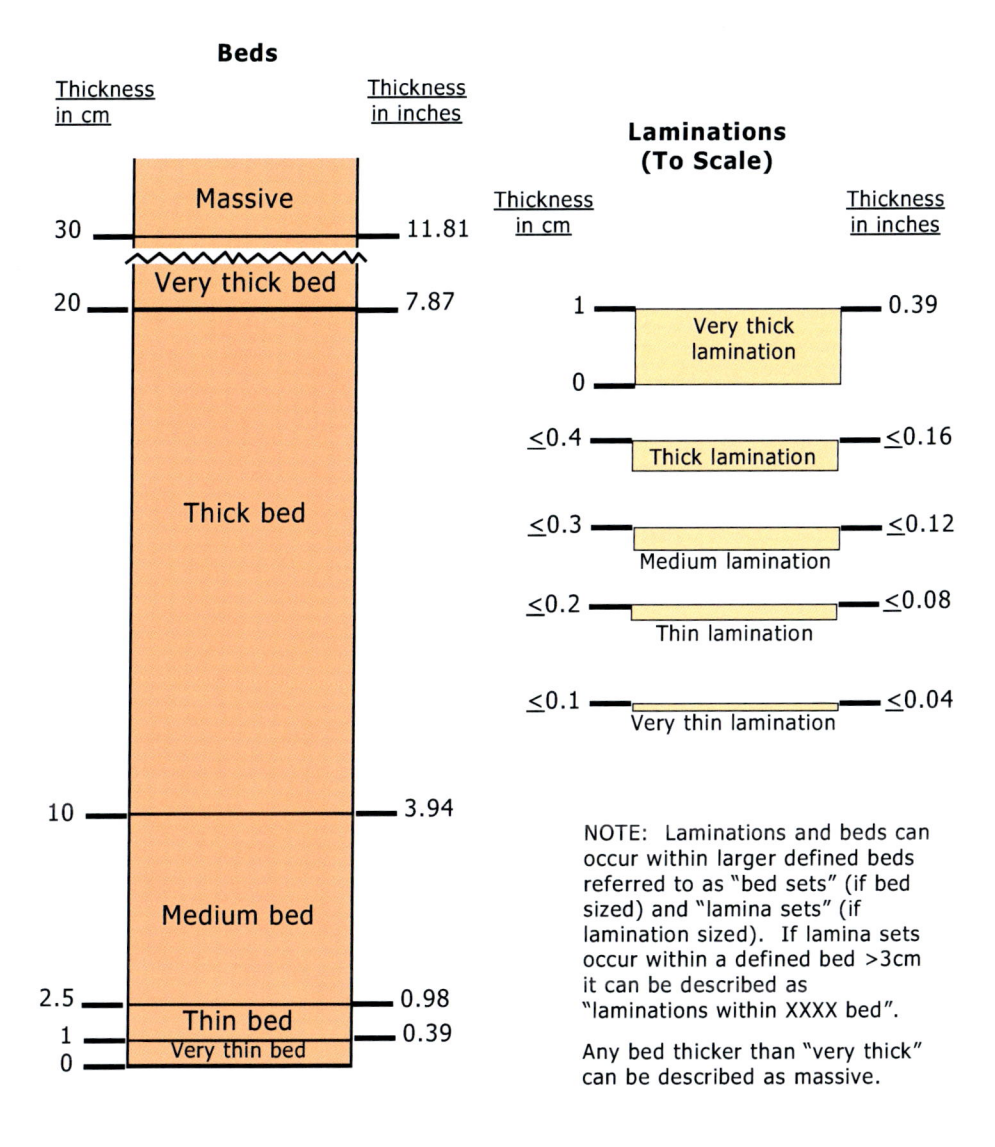

Beds

Thickness in cm		Thickness in inches
30	Massive	11.81
20	Very thick bed	7.87
	Thick bed	
10		3.94
	Medium bed	
2.5		0.98
1	Thin bed	0.39
0	Very thin bed	

Laminations (To Scale)

Thickness in cm		Thickness in inches
1 to 0	Very thick lamination	0.39
≤0.4	Thick lamination	≤0.16
≤0.3	Medium lamination	≤0.12
≤0.2	Thin lamination	≤0.08
≤0.1	Very thin lamination	≤0.04

NOTE: Laminations and beds can occur within larger defined beds referred to as "bed sets" (if bed sized) and "lamina sets" (if lamination sized). If lamina sets occur within a defined bed >3cm it can be described as "laminations within XXXX bed".

Any bed thicker than "very thick" can be described as massive.

Color and Minerology

Although color cannot be used to identify a rock, in sedimentary rocks, it can be used to indicate the presence of other minerals and/or elements. Although this is a general guide for color in sedimentary rocks, but it isn't always 100%.

Most sedimentary rocks that consist of just their base mineralogy will be **gray blue**, **gray** or a **light brown**.

Reds and **orange** shades indicate the presence of iron minerals. So does **yellow**, but yellow is specific to limonite.

Purple shades can also indicate the presence of iron, but manganese will also give sedimentary rocks a purplish hue.

Dark brown and **black** is usually indicative of organic carbon.

Dark green and **green gray** shades indicate the presence of the mineral glauconite, which is only known to deposit in a marine environment.

Olivine and/or pyroxene can also make some sedimentary rocks look dark green. But these are extremely rare, usually near active volcanic islands, and there are no known examples in the Midwest of olivine rich sedimentary rocks.

Dark gray can indicate the presence of organics in things like shale. In things like BIF, it represents unweathered iron minerals.

INDURATION

In most engineering and environmental applications, this is usually the first thing mentioned. Although this could have been divided up into specific rock types, I decided to include it here to avoid extra pages. Clastic rocks e.g. sandstones, conglomerates, and mudstones (siltstones and shales) have different relative hardness terms than all other rocks. This is a qualitative scale. There's no quantitative version for geologic purposes.

Relative Hardness Chart

Sandstone and Conglomerate	Mudstones	Non Clastic Rocks	Description
Unconsolidated	Soft	Poorly indurated	Breaks apart in your hand with very little pressure.
Very friable	Very fissile	Very weak	Breaks apart in your hand with some resistance.
Friable	Fissile	Weak	Breaks apart in your hand with great effort.
Indurated	Somewhat indurated	Somewhat indurated	Does not break in hand but will break easily with a hammer.
Moderately hard	Moderately indurated	Moderately indurated	Breaks with a strong hit with a hammer. Sometimes small pieces can be broken off with a knife.
Hard	Well indurated	Well indurated	Breaks apart with a hammer after significant and repeated hits.
Extremely hard	Extremely well indurated	Extremely well indurated	Only small flake are broken off or small are made after repeated and hard hits with a hammer.

COMMON SEDIMENTRY MINERALS

Remember, when running a hardness test, you have to make sure that you are pressing into the mineral only. Do not cross over to other minerals. If crystals aren't large enough to be scraped, then you cannot perform a hardness test. When conducting the test, use a significant bit of pressure. Don't just lightly drag the tool over the mineral. That being said, don't press so hard that you are trying to dig in, you could break the tool. Below and on the next page, are charts of common mineral hardness in igneous rocks. **Bold** text = defining mineral. Red + **bold**, if defining mineral does *NOT* usually occur in sedimentary rock. This list is in no way comprehensive.

MINERAL	HARDNESS RANGE	STREAK	OTHER CHARACTERISTICS
Ankerite	3.5-4	White	———
Apatite	5	——-	Exists only as detrital grains in clastic sedimentary rocks
Azurite	3.5-4	Light blue	———
Biotite	2.5-3	White	Occurs as angular lithic mica flakes in clastic rocks
Calcite	3	White	Reacts strongly with cold dilute HCl
Copper (native element)	2.5-3	Shinny orange red	———
Corundum	9	White	Often occurs as detrital grains in clastic rocks and carbonates. Highly resistant to weathering.
Dolomite	3.5-4	White	Reacts weakly with cold dilute HCl
Feldspars (general)	6-6.5	White, rarely light gray	Will often appear red and yellow in sandstones. Striations on grains are common.
Fluorite	4	White	Can very rarely be fluorescent

MINERAL	HARDNESS RANGE	STREAK	OTHER CHARACTERISTICS
Goethite (a.k.a. brown ochre)	5-5.5	Brown to yellow brown	Very weakly magnetic
Gypsum	2	White	Dissolves in hot dilute HCl
Hematite	5-6	Bright red to bright dark red	Most common iron oxide in BIF. Magnetic if heated to the Curie Point (769.9°C).
Ice (0°C and 1 atm)	1.5	White	Old and cold ice can have a hardness up to 6
Iron (native element)	4-5	Gray	——————
Lead (native element)	1.5	Shinny light gray	——————
Limonite	4-5.5	Yellow brown	Tends to exist only as a coating on weathered rock
Magnesite	3.5-4.5	White	Can fluoresce pale green to pale blue
Magnetite	5.5-6	Black	Magnetic (even weak magnets will stick to it)
Malachite	3.5-4	Light green	——————
Mica (general)	1-6	White	Occurs as angular lithic mica flakes in clastic rocks
Muscovite	2-2.5	White	Occurs as angular lithic mica flakes in clastic rocks
Orthoclase (plagioclase)	6	White	Exists only as detrital grains in clastic sedimentary rocks

MINERAL	HARDNESS RANGE	STREAK	OTHER CHARACTERISTICS
Pyrite	6-6.5	Black green to black brown	—————
Quartz	7	White	Exists very commonly as detrital grains in sedimentary rocks
Siderite	3.75-4.25	White	Specific gravity = 3.96
Talc	1	White to pearly gray	Fluoresces orange in short UV and yellow in long UV
Topaz	8	Non to White	—————
Uraninite (a.k.a. pitchblende) UO_2 oxidizes to U_3O_8	5-6	Very dark brown or olive green	Isometric crystals. It's radioactive. Density = 6.5-10.95g/cm^3. Ave. density = 8.72g/cm^3. Common in as detrital grains in lower Huronian (pre-GOE) coarse sediments.
Zircon	6-7.5	White	Exists only as detrital grains in clastic sedimentary rocks. It's radioactive.

SEDIMENTARY ROCK TYPES BASED OFF TRADITIONAL GROUIPING

You should *not* classify rocks based off the chart below. I am just showing you this chart because you will often see rocks grouped in this manner even though it is interpretive. Yes, I do realize that calling a rock made of sea shells "biological" isn't much of a stretch, but we usually classify rocks based off mineralogy and not origin.

The top three clastic, chemical, and biological are interpretations, they are not based off lithology. Although lithology is used below the top three. That's why you see carbonates under chemical and biological and chert is under all three! It is this type of "one in all or all in one" interpretive classification, that I am trying to avoid by using just lithology.

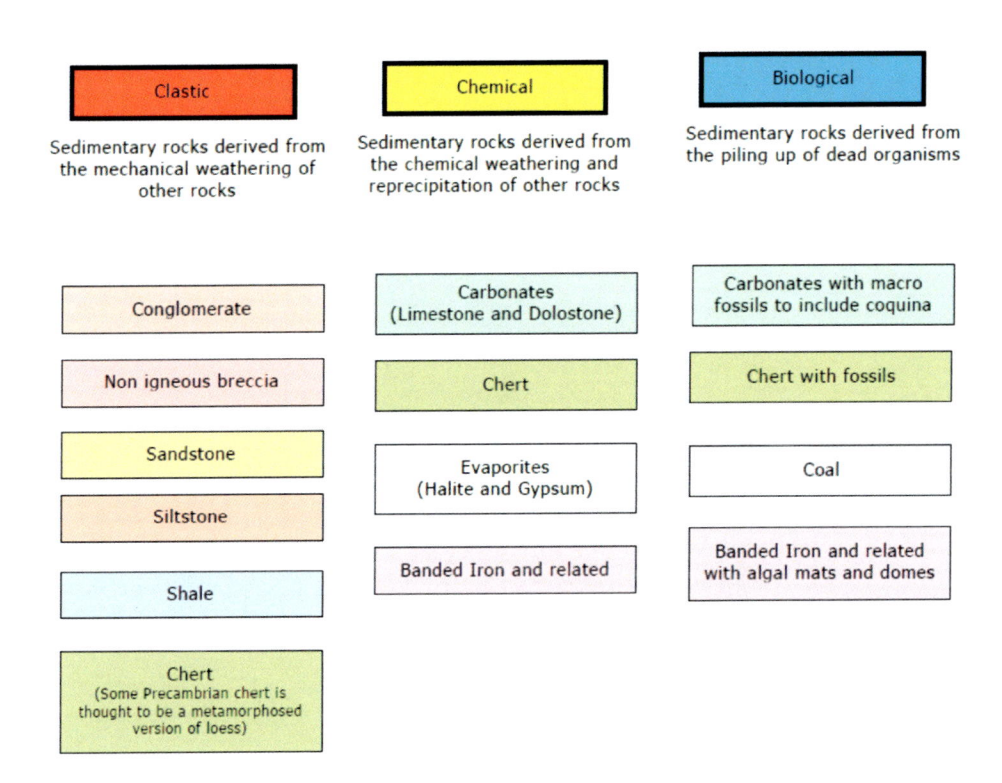

Clastic
Sedimentary rocks derived from the mechanical weathering of other rocks

- Conglomerate
- Non igneous breccia
- Sandstone
- Siltstone
- Shale
- Chert (Some Precambrian chert is thought to be a metamorphosed version of loess)

Chemical
Sedimentary rocks derived from the chemical weathering and reprecipitation of other rocks

- Carbonates (Limestone and Dolostone)
- Chert
- Evaporites (Halite and Gypsum)
- Banded Iron and related

Biological
Sedimentary rocks derived from the piling up of dead organisms

- Carbonates with macro fossils to include coquina
- Chert with fossils
- Coal
- Banded Iron and related with algal mats and domes

PART II
CLASTIC SEDIMENTARY ROCKS

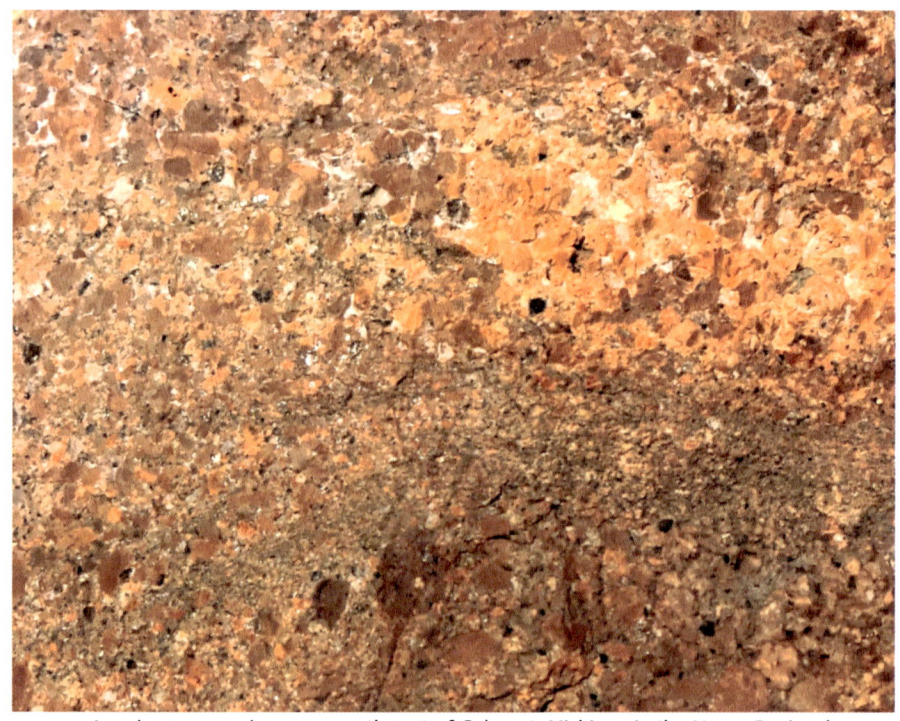

Angular paraconglomerate northwest of Calumet, Michigan in the Upper Peninsula.

WHAT ARE CLASTIC ROCKS?

A rock is a solid that is an aggregate of one or more minerals, or a solid body of undifferentiated mineral matter (USGS, 2015). Rock units can be either sediments or hard rock. There is no induration (a.k.a. hardness) requirement for the status of what is considered a rock unit. The reason being, is that rock units can have highly variable indurations both laterally and vertically. That being said, a rock in a box usually has a specific hardness to it that can be calculated from its minerology. I have mineral and rock hardness tests on p. 190.

Clastic rocks are inorganic sedimentary rocks that are made of pieces of other rocks.

This excludes fossil rocks such as coal and peat. It also excludes all carbonates, regardless of origin, and all rocks directly precipitated from bodies of water (evaporites).

The term "clastic" isn't in the rock name. Usually, neither are one of the three types, conglomerates, sandstones, or mudstones. Although sedimentary breccia and diamicton do exist. The three main types are divided as below:

      <u>Conglomerate</u>

            Paraconglomerate
            Orthoconglomerate
            Dioconglomerate

      <u>Sandstone</u>

            Arenite
            Wacke
            Greywacke

      <u>Mudstone</u>

            Siltstone
            Shale

CONGLOMERATE

Conglomerates are the easiest rocks to classify because the grains are so large (>50% at >2mm in size). The term breccia, like with debris flows, is often used for sedimentary rocks. I avoid doing this. Sedimentary breccia can be described as angular conglomerates. There are only 3 parts to the naming part of a conglomerate. You have the "subtype" followed by the "matrix modifier" if applicable (see chart on the next page), and finally your "main type".

You can expand on you descriptions after naming the conglomerate. You can add things like the lithologies of the pebbles.

Paraconglomerate, Orthoconglomerate, and Dioconglomerate

Conglomerates come in three "main types". This will be the very last word of the name. "Paraconglomerate" is matrix supported; as is common with some diamictons and debris flows. "Orthoconglomerate" is clast supported; as is the case with river bars, some alluvial fans, outwash, and some desert pavement. "Dioconglomerate" are matrix and clast supported; as is common with bedded conglomerates like fluvial streams and some alluvial fans. Paraconglomerate, orthoconglomerate, and dioconglomerate will be your rock type.

Oligometic and Polyometic

There are two "subtypes" that can be added to the rock type you can choose, or leave off if you don't need a detailed description. "Oligometic" is for conglomerates who's clasts are $\geq$80% one rock type. "Polyometic" is used if there are clasts of <80% one rock type or multiple and varying lithologies. Subtype is always listed as the first part of the name.

Conglomerates with Matrix

If a conglomerate has significant matrix (a.k.a. clasts finer than pebbles), the following chart is to be used and the appropriate modifier will be added in between the subtype and the type. The chart below is to be used for conglomerates with significant matrix, and is different from previous editions and is introduced herein. The chart gives you your matrix modifier if applicable.

(2nd order normalization)

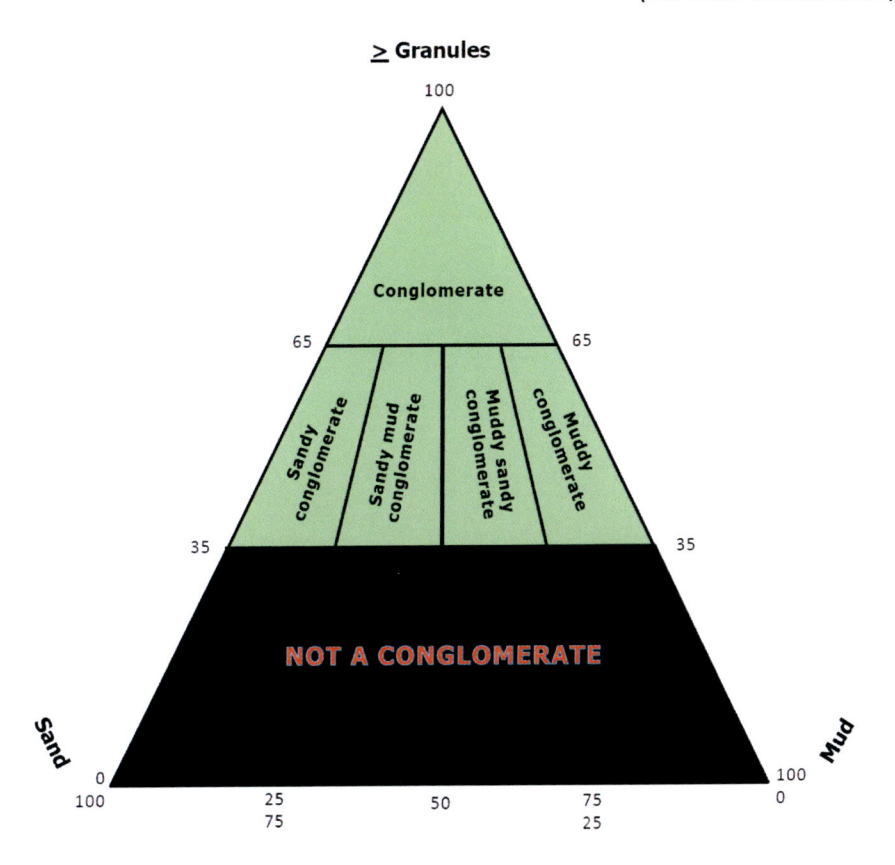

≥ **Granules**
100

Conglomerate

65 | 65

Sandy conglomerate | Sandy mud conglomerate | Muddy sandy conglomerate | Muddy conglomerate

35 | 35

NOT A CONGLOMERATE

Sand — Mud

0 — 100
100 — 0
25 / 75
50
75 / 25

NOTES: *All numbers are in percentages. This chart functions like a normal ternary plot. The word "conglomerate" would be substituted for the appropriate type, i.e. paraconglomerate, orthoconglomerate, or dioconglomerate.*

Adapted from: Baumann and Michaels (2011), and Folk (1954)

Using Cores for Training

This may look like conglomerate, but it is concrete. By definition, rocks are natural. That doesn't mean you can't use things like this to teach others about color, texture, reactivity, composition, and matrix.

Lithological Conglomerate Symbols

It is often easier to graphically depict a certain rock type in either field notes or a lab book. Below are lithological symbols for conglomerates.

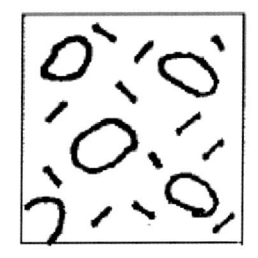

Matrix supported conglomerate in a mud dominated matrix

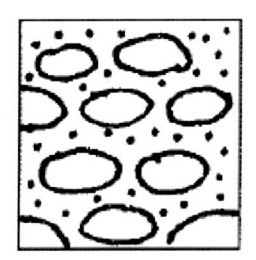

Clast supported conglomerate with no matrix or sand dominated matrix

Conglomerate with calcite veins

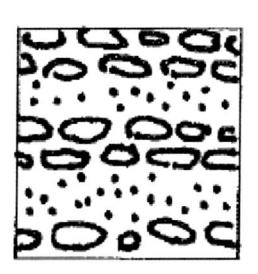

Bedded conglomerate

SANDSTONES

Calling a clastic sedimentary rock a sandstone has nothing to do with composition and everything to do with grain size. A sandstone has particles that fall between 0.062mm and 2mm in size, just like a sand (p.173-174). Sandstones can be further divided based on composition. Relative hardness for sandstones is on p. 185.

Sandstone Maturity and Lithology

We use maturity to generically refer to the quartz richness of a sandstone. Sandstones are divided into 3 main parts. Lithics, feldspars, and quartz. An immature sandstone has <70% quartz. A mature sandstone has grains $\geq$70% to 90% quartz. A super mature sandstone has >90% quartz grains.

Maturity of a sandstone isn't sufficient if you need detailed analysis. So lithology can be used. You will have two "main types". There are "arenites" (sandstones with $\leq$15% fines), and "wackes" (>15% to ~50% fines). Once that is determined you use ternary plots to classify your sandstone based on the relative amounts of lithics, feldspars, and quartz grains. These percentages are plotted on a "quartz, feldspar, lithic" (QFL) ternary diagram.

Quartz grains are any grains composed of actual quartz. This is actually the most common mineral in sandstones. The individual quartz grains are usually colorless to white with some translucence. I personally do *NOT* include things like chert with quartz. Chert is included in lithics.

Feldspar grains include all feldspars (alkali feldspar and plagioclase), since the smaller the grains the less likely you are to see striations. They are usually (but not always) opaque and white, yellow, pink, or light brown in color.

Lithics include everything else. This can be fragments of carbonates, volcanics, metamorphic rocks, other sedimentary rocks, and heavy minerals.

QFL Arenites

Arenites are sandstones with ≤15% matrix (a.k.a. fines).

An arenite would be plotted on the ternary diagram below similar to a QAPF plot. After normalizing for quartz, feldspar, and lithics by removing any accessory minerals and cement.

(2nd order normalization)

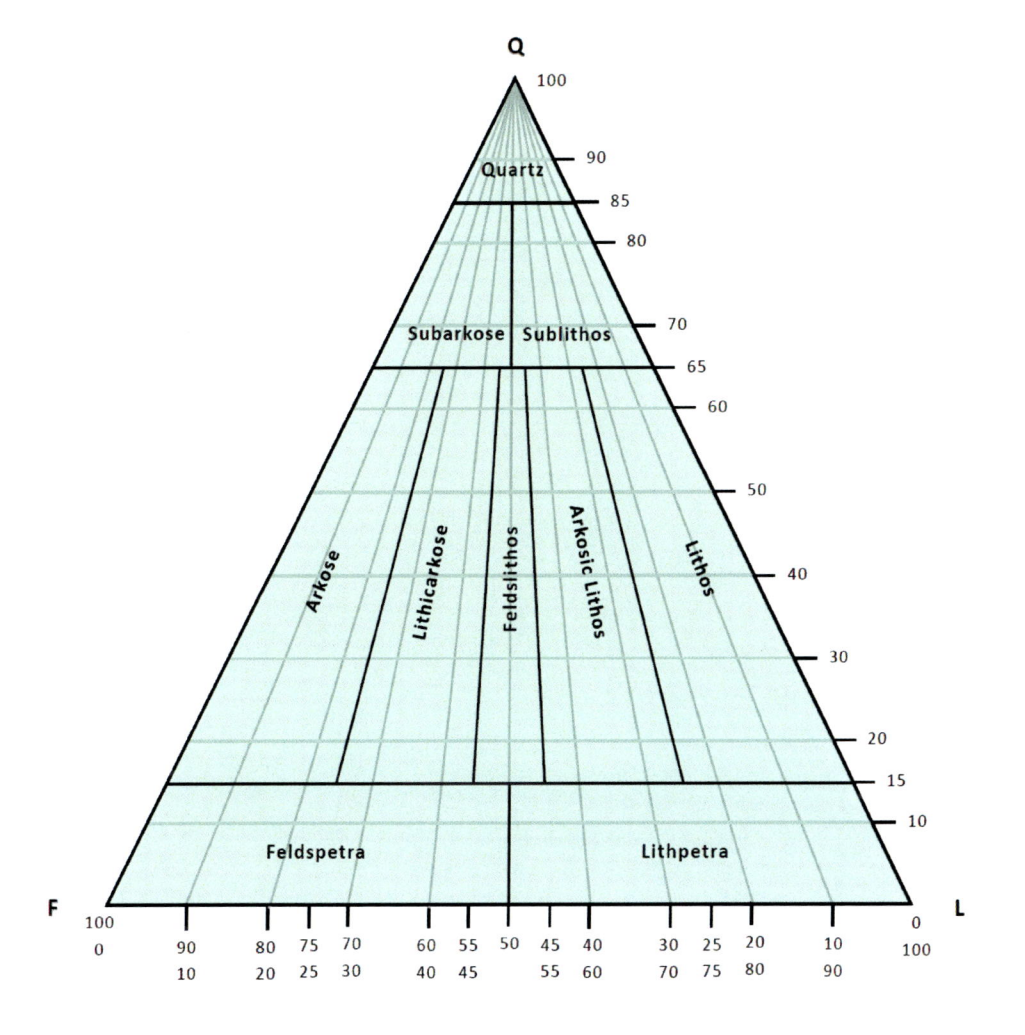

Lith is Greek for stone
Petra is Latin for rock

Quartz includes monocrystalline and polycrystalline. Chert is put in with lithics and excluded from quartz as in Folk et al. (1970).

Specimens with less than 15% fines and matrix have "arenite" after the first word. If there is more than 15% up to 50±5% the word "wacke" is added to the end of the name.

QFL Wackes

Wackes are sandstones with >15% to 50±5% matrix (a.k.a. fines). Why the ±5%? There is a bit of a discretion here. You had only 47% sand but it was mostly coarse; then you would likely call it a wacke as opposed to a mudstone. If you had 55% sand grains but they are all very fine to fine, you may want to call it a mudstone instead of a sandstone.

A wacke would be plotted on the ternary diagram below similar to a QAPF plot. After normalizing for quartz, feldspar, and lithics by removing any accessory minerals and cement.

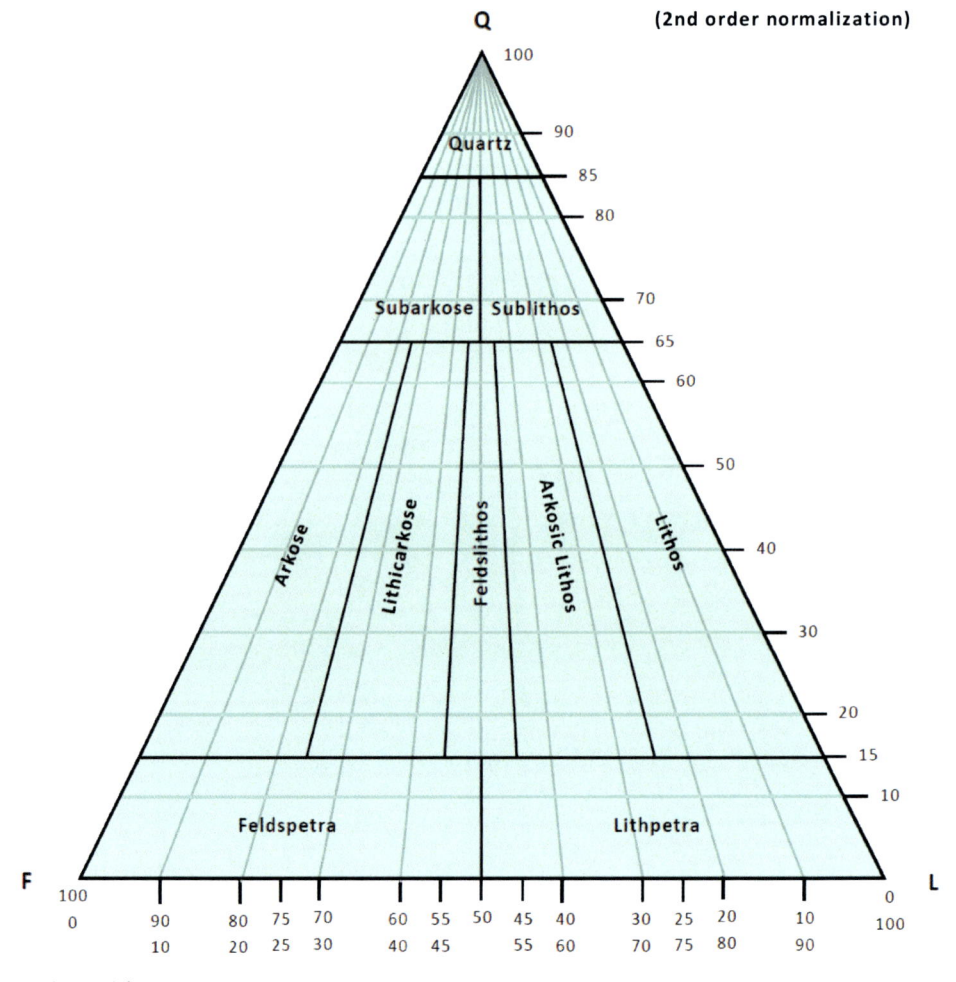

Lith is Greek for stone
Petra is Latin for rock

Quartz includes monocrystalline and polycrystalline. Chert is put in with lithics and excluded from quartz as in Folk et al. (1970).

Specimens with less than 15% fines and matrix have "arenite" after the first word. If there is more than 15% up to 50±5% the word "wacke" is added to the end of the name.

Maturity/Texture 3D Diagram

This is a 3D chart to illustrate how the QFL arenite and wacke charts relate to maturity and texture. In this case texture is the relative amount of fines.

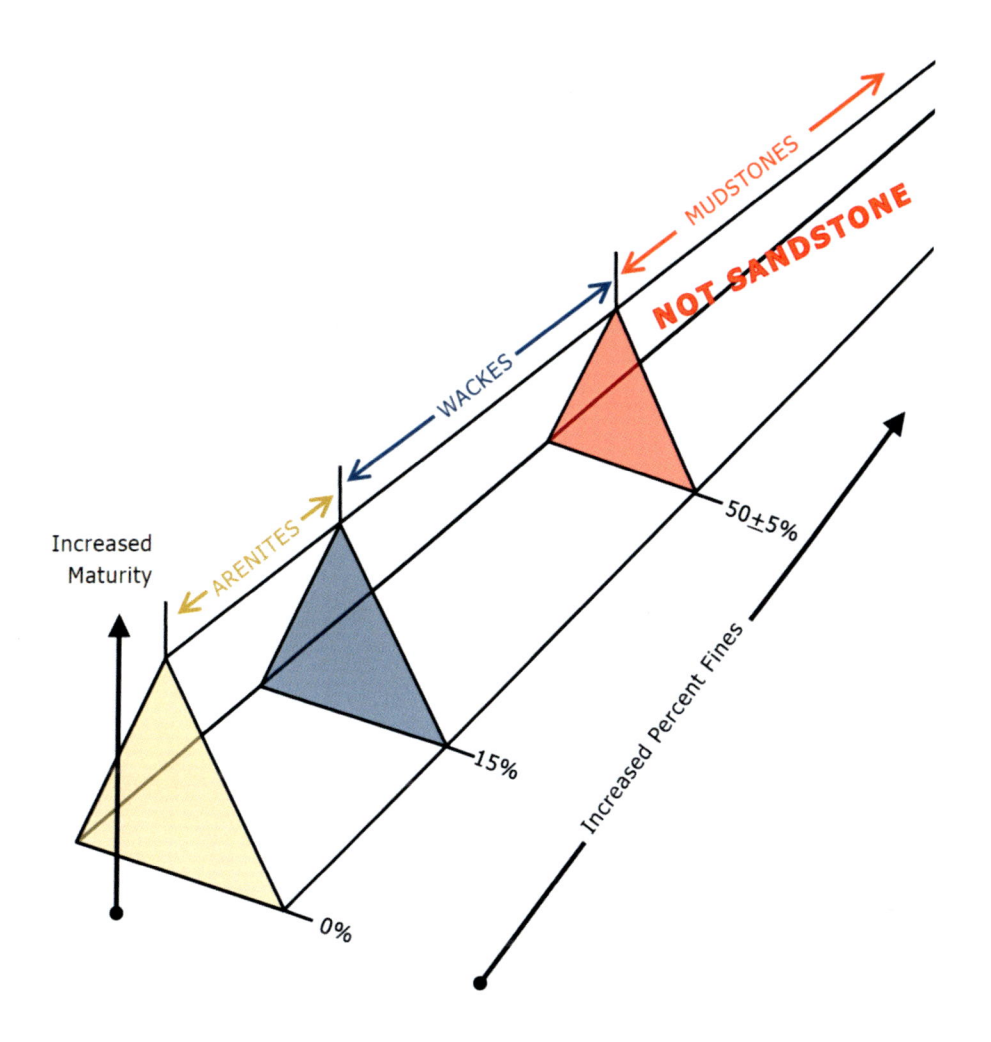

Adapted from: Folk (1954) and Greensmith (1989)

Comprehending the reason why 50/50 isn't the standard line

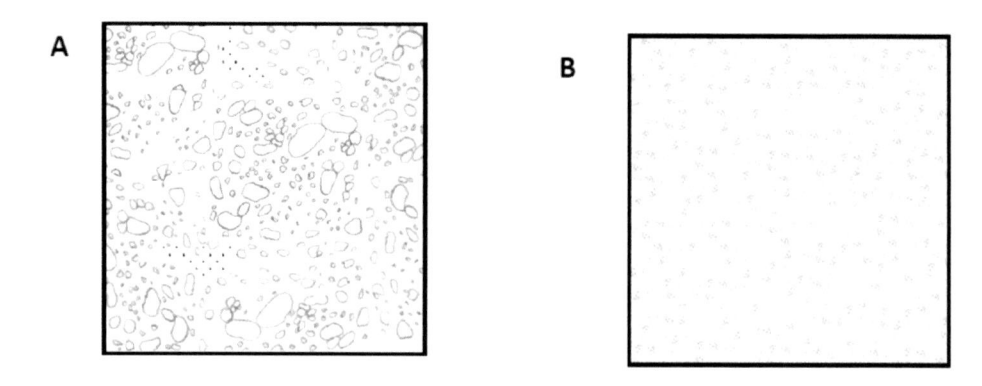

A

B

Both diagrams contain around 50% sand or coarser particles. Silt and clay are the blank or very light gray areas between the grains. A) Should be classified as a sandstone due to the amount of coarse sand and granules. But it is a wacke, not an arenite. B) Contains a lot of sand but it is all very fine to fine so it would be better to call it a sandy mudstone, and not a sandstone.

Greywacke

Greywacke (or graywacke) is a dying term, but it still frequently appears in literature. The sedimentary rocks that are considered to be greywacke's are extensive in the Midwest (e.g. Michigamme, Tyler, and Rove Formations), especially around Lake Superior. It predates the QFL diagrams by at least 125 years. McBride (1963) was the first to generate a ternary diagram for sandstones. Although the term maybe older the oldest printed definition I could find, which is in Andrew Ure's (1840) "Dictionary of Arts, Manufactures, and Mines" on page 619 it defines it as, "*GRAUWACKE or GREYWACKE, is a rock formation, composed of pieces of quartz, flinty slate, felspar and clay slate, cemented by a clay-slate basis; the pieces varying in size from small grains to a hen's egg.*" A modified definition appears in Ure's 1878 volume II redone version on pages 736-737. Volume I (the original 1840 print) was reprinted in 1870. That definition predates The Wentworth scale by 82 years.

A greywacke is little more than a poorly sorted dark colored wacke. It really has no modern use with QFL diagrams. Most "greywacke" would plot on the QFL wacke diagram as a "lithic wacke, feldspathic wacke, lithic arkosic wacke, or arkosic wacke". It's a very ad hoc term. Some usages would even force a plot on the QFL arenite diagram. Sometimes it's even used for mudstone and rarely even conglomerates. So there is no need or purpose for the continuation of the word, and its usage has slowly been falling out of favor.

You should avoid using the term greywacke, but you need to be aware of it.

Bouma (Turbidity) Cycle Sequence

A Bouma cycle is a special type of bedding usually associated with greywackes. That's why it is included here and not with bedding. A Bouma Cycle (or turbidity cycle) is repeating set of deposits in a low energy, clastic, underwater flow (finer grained sediments). Individual layers do not have set thicknesses. Layers "E" and D" are often eroded and incorporated into a new overlying Layer "A". "E" and "D" are often the thinnest of the layers. "E" tends to be massive (lacking internal structure) mud. "D" tends to be laminated mudstone. "C" is rippled and usually a fine grained sandstone wacke. "B" tends to be planar laminated, sorted, medium grained sandstone. "A" is usually the second thickest (although it can be the thickest) unit in the cycle. "A" is typically comprised of massive medium to coarse grained sandstone. Sometimes it will contain stringers of mud.

These types of deposits are very common in the forearc basin deposits that accumulated during the Penokean orogeny and in some deposits of the Huronian Supergroup in western Ontario, Minnesota, northern Wisconsin, and the Upper Peninsula of Michigan. The formations that exhibit these cycles the most are the Michigamme (Upper Peninsula), the Tyler (northern Wisconsin), and the Rove Formations (Minnesota and western Ontario).

Cross sectional view of a typical Bouma Cycle

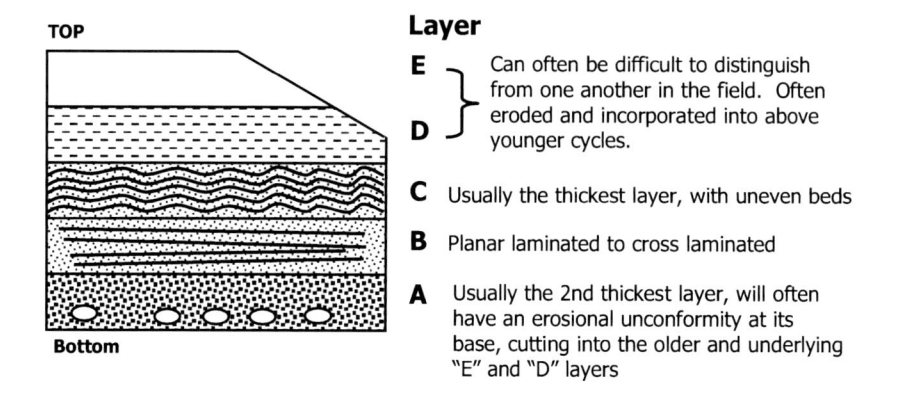

Layer

E, D — Can often be difficult to distinguish from one another in the field. Often eroded and incorporated into above younger cycles.

C — Usually the thickest layer, with uneven beds

B — Planar laminated to cross laminated

A — Usually the 2nd thickest layer, will often have an erosional unconformity at its base, cutting into the older and underlying "E" and "D" layers

Adapted from: Bouma and Ravenne (2004)

Lithological Sandstone Symbols

It is often easier to graphically depict a certain rock type in either field notes or a lab book. Below are lithological symbols for sandstones.

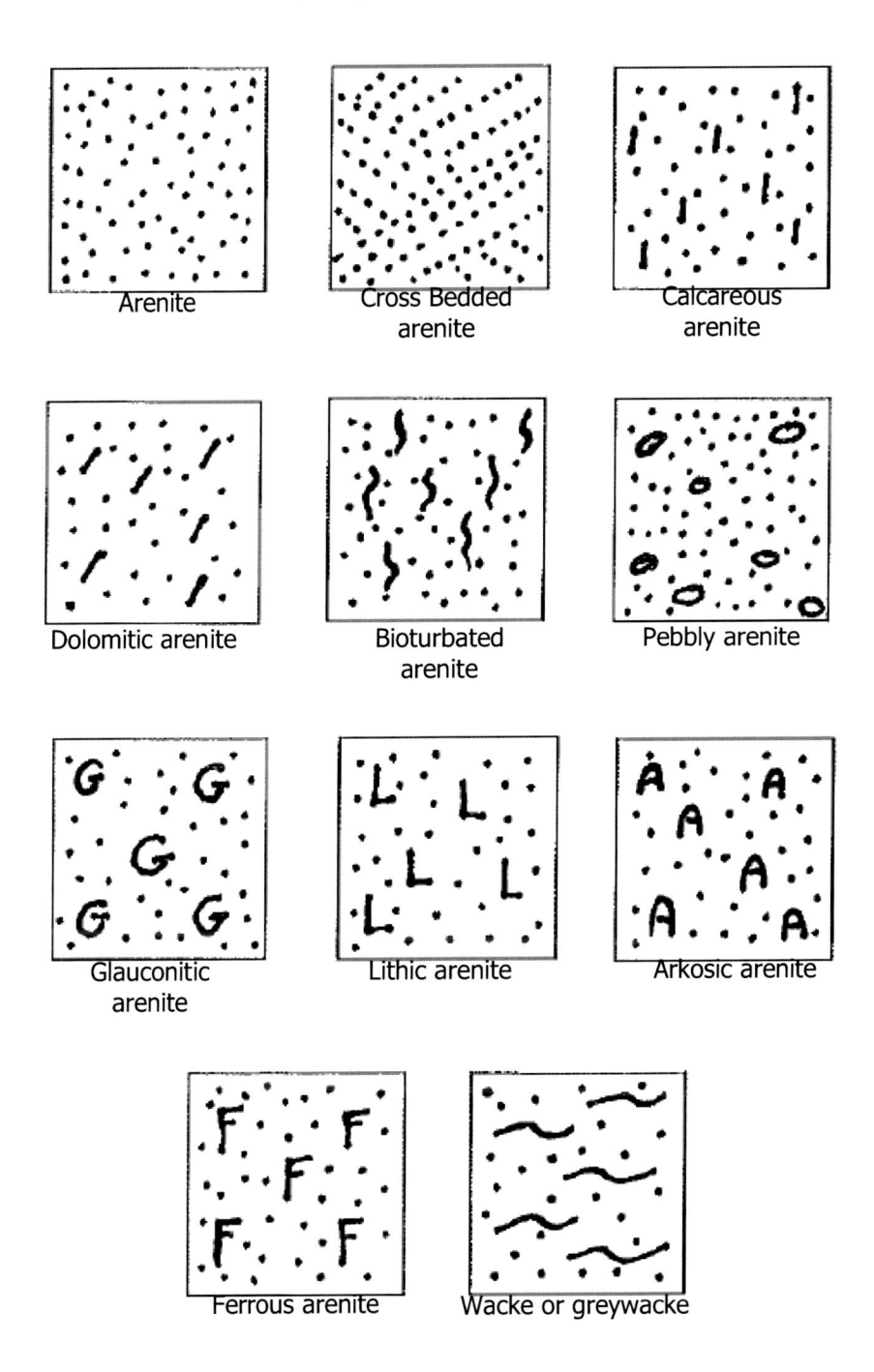

Arenite

Cross Bedded arenite

Calcareous arenite

Dolomitic arenite

Bioturbated arenite

Pebbly arenite

Glauconitic arenite

Lithic arenite

Arkosic arenite

Ferrous arenite

Wacke or greywacke

MUDSTONE

Mudstones are perhaps the simplest of all sedimentary rocks to classify due to such a small grain size. Even with magnification, under a conventional microscope, it is impossible to see most clay minerals. Silt sized grains can possibly be made out, if the silt is mostly coarse. For this reason mudstones are essentially divided into two types; siltstone and shale. If there is about 50% silt and 50% clay, it's generally just referred to as a mudstone.

Siltstone

Siltstones are somewhat rare in marine settings, but common in fluvial and alluvial environments. Like sandstone they are defined by being mostly composed of silt grains between 0.004mm to 0.062mm in size. They tend to have abundant sand and clay in them. If there is a significant amount of clay sized grains in the siltstone, the word "clayey" is *NOT* used. The word "argillaceous" is (see next page).

Shale

Shale is composed of mostly clay sized grains <0.004mm, for classification purposes (p. 173). There is some overlap between silt and clay but for visual classification purposes the line is set at 0.004mm. There's no way you are going to see a gain size smaller than that anyway without either a really powerful optical microscope (the smallest thing visible is about 0.003mm or 3 micrometers), or a scanning electron microscope.

Modifiers

Modifiers in mudstones are minor inclusions of other minerals or structural features. These are far more useful than grain size in telling apart mudstones. Modifiers would be listed before the rock type if they are significant. Some common mineralogical ones are cherty, hematitic, glauconitic, micaceous, and calcareous. Common structural modifiers are, very thinly laminated, even cross laminated, bioturbated, pitted, or varved. Modifiers are added before the "type" (a.k.a. name).

Type of Mudstones

This chart is pretty much as good as you can get in identifying a mudstone without powerful microscopes.

Name	Definition
SHALE	>50% of the grains are clay to fine silt sized
SILTSTONE	>50% of the grains are fine to coarse silt sized
SANDY SHALE	35% to 50% of the grains are very fine to very coarse sand sized, with almost no silt, the rest is clay sized
SILTY SHALE	35% to 50% of the grains are fine to coarse silt sized, with almost no sand, the rest is clay sized
SANDY SHALE with SILT	35% to 50% of the grains are very fine to very coarse sand sized, with 5% to <35% silt, the rest is clay sized
SILTY SHALE with SAND	35% to 50% of the grains are fine to coarse silt sized, with 5% to <35% very fine to very coarse sand, the rest is clay sized
ARGILLACEOUS SILTSTONE	35% to 50% of the grains are clay sized, with almost no sand, the rest is silt sized
SANDY SILTSTONE	35% to 50% of the grains are very fine to very coarse sand sized, with almost no clay, the rest is silt sized
SANDY SILTSTONE with CLAY	35% to 50% of the grains are very fine to very coarse sand sized, with 5% to <35% clay, the rest is silt sized
ARGILLACEOUS SILTSTONE with SAND	35% to 50% of the grains are clay sized, with 5% to <35% very fine to coarse sand sized, the rest is silt sized

NOTE: *You may be wondering why some of these percentages add up to more than the first listed 50% requirement. It's due to the variability and relative ratios of the constituents that are <50%.*

Clay Minerals

Clay is very common on the surface and forms the bulk of most mudstones. Clay is not mechanically broken down rock like ≥silt sized grains are. They form from the chemical alteration of other minerals, mostly of micas and feldspars due to the presence of liquid water at the surface. Due to their extreme small size, individual minerals won't be named. There are four main groups, chlorite clay is rare and its properties are difficult to differentiate macroscopically from kaolin and illite. Chlorite clays are not included below or on the next page. The cation exchange for chlorite clays is about 30-50meq/100g.

Illite Group

Illite is the most common group of clay in the Midwest and forms the majority of the clay in clay rich Quaternary diamicton. Its shrink-swell capacity is very low. It's very soft (Mohs 1-2). It is usually light gray or light blue gray to gray in color. Gray, green, and reddish hues are also common. Its cation-exchange capacity is less than smectite but higher than kaolin at 20-30meq/100g. Its type is the Maquoketa Shale in Illinois, where illite gets its name.

Kaolin Group

Kaolin clays are the second most abundant clays in the Midwest. They are usually very light if not white in color; usually pale yellows and oranges. They are soft (Mohs 2-2.5). They have low shrink-swell and low cation-exchange capacity at 1-15meq/100g.

Smectite Group

The smectite Group herein includes all clays with high shrink-swell and high cation-exchange capacity (25-100meq/100g). This includes bentonite. Smectite clays are most common in glacial bog varve and glacial lacustral deposits. Bentonite occurs in the Midwest mostly as thin yellow to white beds in Phanerozoic rocks. Its color (to exclude bentonite) is usually light gray to dark gray. It is the rarest of the groups. Its specific gravity is low at 1.7-2.0. It has a hardness of <2 (Mohs).

Testing Clay Minerals

Although not usually a part of identifying rocks, testing for relative amounts of illite, kaolin, and smectite can aid in the differentiation of facies within shales.

In order to conduct these tests, the shale has to be nearly free of sand (<5%) and silt (<10%) by mass. A lot isn't needed as this will be a volume test, 50-100 grams should be enough in most cases. The shale must be mechanically broken down into a fine powder and allowed to air dry for ≥ 24 hours. Once air dried, add the powder to a small cylinder of known volume. Lightly tap the cylinder on the table (to slightly compact the powder) and draw a line on the cylinder at the top of the leveled powder. Calculate the initial volume (V_o). Add distilled water until the powder is covered at least up to twice the powder volume before adding the water and shake it so all the powder can absorb water. Let it sit for ≥ 36 hours undisturbed. Now draw a line to where the top of the powder is in solution (V_1). Calculate the difference (in percent) between V_o and V_1. Use the chart below to gauge relative clay groups.

Change in Volume of Select Clay Mineral Groups

Change in Volume (% Increase)	Mineral Group
0 to 5	Illite with minor kaolin
≥ 5 to 10	Illite and Kaolin
≥ 10 to 15	Kaolin with minor illite and/or smectite
≥ 15 to 25	Kaolin and Smectite
≥ 25 to 35	Smectite with minor kaolin
≥ 35-50	Smectite
≥ 50	Smectite (consisting of bentonite)

NOTE: *It may be possible to calculate a small decrease in volume (a negative percent) between V_o and V_1. This occurs because the volume of powder recorded before you add water is loosely packed. If this occurs, there are no swelling clays and the powder is all illite.*

Diamicton

In the Midwest, diamicton essentially refers to glacial till. The term was introduced by Flint, et al. (1960) as a substitute for the barely used term symmictite. We have been abandoning the term till slowly since the 1990s. Hansel and Johnson (1996) were the first to directly apply the term diamicton, and it has been gaining acceptance ever since. Diamicton was introduced in an effort to eliminate descriptions of lithological units that indicate an origin.

Diamicton is essentially, "an unsorted to poorly sorted sediment that contains particle sizes from clay to boulders, that are dispersed (or suspended) in a finer matrix (i.e. matrix supported), that lacks any easily discernable internal structure (e.g. bedding, ripple marks, and preferred orientation of grains)".

Although this pretty much describes all glacial till, it also described certain debris flow deposits like the Aweres Formation (Ontario) and could be used to describe parts of the Baldwin Conglomerate (Wisconsin).

Diamicton is the sediment term. The term diamictite is used for the rock name. All Quaternary diamicton is unconsolidated. There is consolidated diamictite within the Huronian Supergroup, like the Coleman Member of the Gowganda Formation. Since the vast majority of diamicton in the Midwest contains suspended particles in a matrix supported to dominated, that is composed nearly always of clay and silt, we can use a modified mudstone chart (see next page).

Paleoproterozoic diamictite of the Coleman Member of the Gowganda Formation in Ontario, Canada. Six inch pencil for scale. The red pebbles and cobbles are granitic rocks from further north-northwest.

Type of Diamictite

This chart is pretty much as good as you can get in identifying a mudstone without powerful microscopes. This chart excludes the pebbles to boulders and normalizes the sand, silt, and clay, which cannot make up >50% of the rock; or you don't have a diamicton.

The suspended clasts (pebbles and larger) have to be addressed. Their dominant lithologies and size are put before the name in the chart below. If a diamicton were 9% limestone cobbles and 7% granite pebbles in a clay with a lot of silt, your diamicton (before normalization) would be named at the minimum:

COBBLE and PEBBLE SILTY DIAMICTON

>15% to 35% ≥pebbles, gets listed first (as in the example above).

5% to 15% ≥pebbles, that gets listed following the name (as shown below) proceeded with the word "with".

The following percentages are based off the total percent within the rock.

Classification of Diamicton

Name	Definition
DIAMICTON	≥50% of the grains are clay sized
SILTY DIAMICTON	35% to 50% of the grains are silt sized in a clay matrix
SANDY DIAMICTON	35% to 50% of the grains are sand sized in a clay matrix
SANDY DIAMICTON with SILT	35% to 50% of the grains are sand sized, with 5% to <35% silt, the rest is clay sized
SILTY DIAMICTON with SAND	35% to 50% of the grains are silt sized, with 5% to <35% sand, the rest is clay sized

NOTE: *The above names are decided once the >pebble sized clasts are removed and the sand-silt-clay is normalized.*

You may be wondering why some of these percentages add up to more than the first listed 50% requirement. It's due to the variability and relative ratios of the constituents that are <50%.

Lithological Mudstone Symbols

It is often easier to graphically depict a certain rock type in either field notes or a lab book. Below are lithological symbols for mudstones.

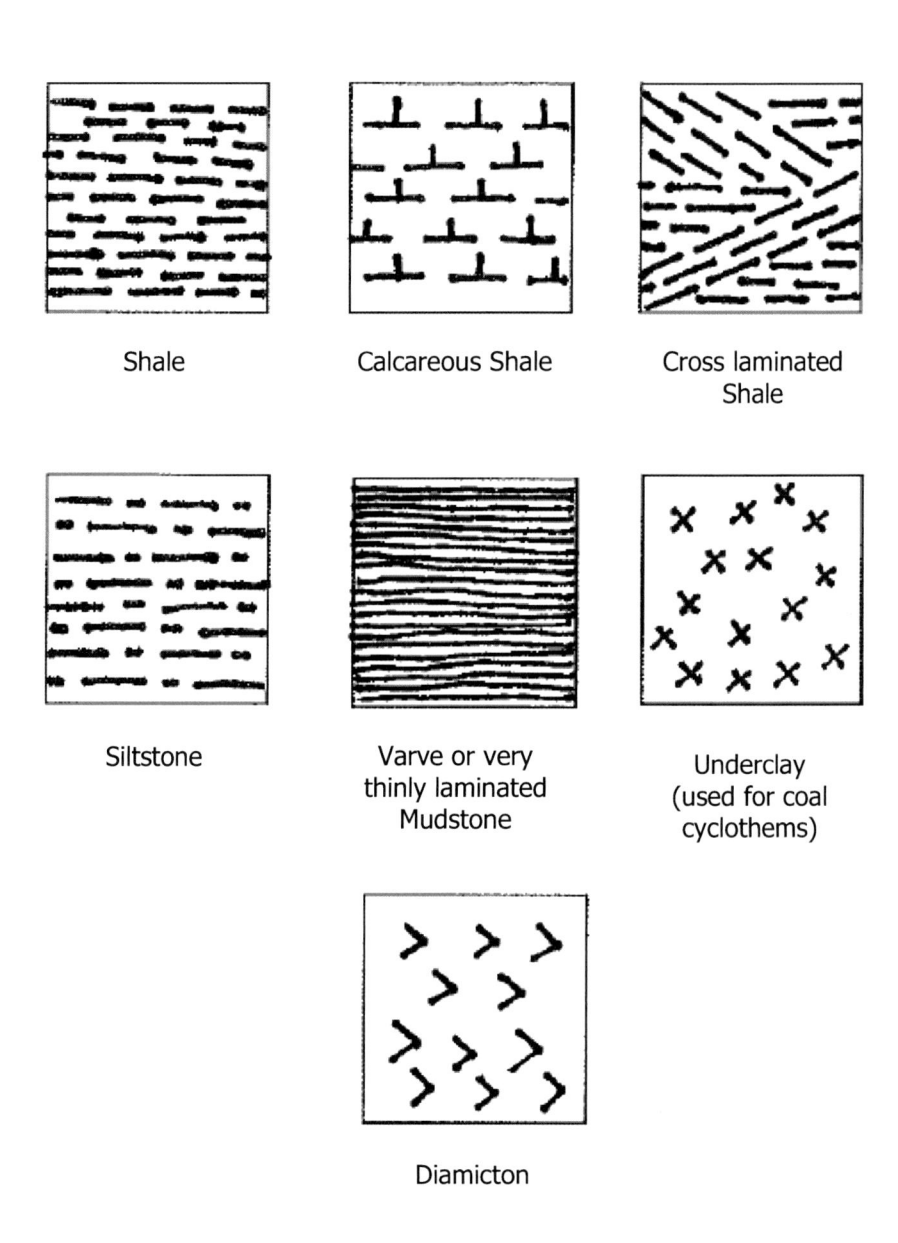

Shale

Calcareous Shale

Cross laminated Shale

Siltstone

Varve or very thinly laminated Mudstone

Underclay (used for coal cyclothems)

Diamicton

PART III
NON CLASTIC SEDIMENTARY ROCKS

Conglomeratic dolostone from near Iron Mountain, Michigan in the Upper Peninsula.

To avoid interpretive classifications, I will just include all other sedimentary rocks as non clastic.

CARBONATES

The most common carbonate rocks in the Midwest are limestone and dolostone. Limestone is mostly composed of calcite $CaCO_3$. Dolostone is mostly composed of dolomite $CaMg(CO_3)_2$. Those two minerals along with lesser amounts of microcrystalline silica (SiO_2) make up the vast majority of carbonate rocks.

Carbonate rocks fall into a gray area. Most rocks easily fall into either clastic, chemical, or organic rock classifications. Sedimentary carbonates do not. Most Precambrian (but not all) carbonate rocks are derived from chemical precipitation of sea water. Many Phanerozoic carbonates are also this way. So they would be chemical. However, many Phanerozoic carbonates boast vast reefs and strata littered with fossils. Although these reefs are composed of carbonates and not organic compounds, they are considered organic or bio carbonates. High concentrations of fossils can lead to complications in classification. Although herein they will be considered grains. This complication is the reason this book is mainly divided into clastic and non clastic rocks.

In order to simplify the classification process, the origin of the carbonate rock will not be taken into account. How do we even identify carbonates in the lab? We conduct an acid/hardness test. The acid standard and carbonate testing flowchart (p. 215) is my own derivation, based on my personal experience. The flow chart shows how to tell if you have a limestone, dolostone, or combination (with or without silica). The flowchart works best for carbonate rocks without clastic inclusions (sand, silt, and clay). Fortunately things like sandy and argillaceous carbonates are rare. It appears that if carbonate production is high, clastic input is low to non existent. If clastic input suddenly increases, then carbonate deposition is shut off. This rule applies to marine carbonates. Lacustral carbonates do not follow this pattern.

For this test to work you need dilute hydrochloric acid (HCl). You can purchase this in the form of muriatic acid in hardware stores and dilute it to the desired percentage yourself.

HCl Acid Standards

Mixing

Muriatic acid is usually ~31% HCl. To make my 3-5% solution, mix 1.5 units of muriatic acid with 8.5 units of distilled water. To make 10-15% solution, mix 4.5 units of muriatic acid with 5.5 units of distilled water. You just add a drop to the part of the rock you want to test.

3-5% solution = Ac ("A" is for acid and "c" for common): 10-15% solution = Ah ("A" is for acid and "h" for higher concentration)

Abbreviations

DS = dolostone, LS = limestone, s = silica or siliceous; e.g. a siliceous dolostone = sDS c = calcareous, d = dolomitic

Reactions

No reaction (Nx) = No visible fizzing and no fizzing heard. As if you just poured water on it.

Weak reaction (Wx) = No visible fizzing but faint fizzing can be heard if your ear is close to the rock.

Moderate reaction (Mx) = Faint visible fizzing and fizzing can be heard from at least 1.5 feet (46cm away). Bubbles may be visible under hand lens.

Strong reaction (Sx) = Fizzing visible and producing bubbles that can be seen easily with a hand lens and to a lesser degree without a hand lens from about 1.5 feet away. Some carbonate rock may be dissolved out where acid was applied; visible in hand lens.

Vigorous reaction (Vx) = Fizzing can be easily seen and heard from greater than 1 yard (~1 meter) away. Some of the rock is easily seen with the naked eye to be dissolved where the acid was dropped.

When testing, begin with Ac. If the rock has clear and different lithologies within, try to keep your drop of acid on the part you are testing.

NOTE: *Mix acids properly with proper personal protective equipment. Never dump acid down the sink. This page serves as the legend for the next page.*

Converting percent acid solution to molarity

To convert percent to molarity (M) you will need the percent of acid in solution (%), density (ρ) or specific gravity (SG) and the molecular weight of the formula (M_W).

$[(\% \times \rho) / M_W] \times 10 = M$

NOTE: Molecular weight (usually unitless) is really just another term for molar mass, which is in grams per mole (g/mol).

The M_W of acetic acid (CH_3COOH)	= 60.05	SG = 1.052
The M_W of citric acid ($C_6H_8O_7$)	= 192.12	SG = 1.661
The M_W of hydrochloric acid (HCl)	= 36.46	SG = 1.185
The M_W of nitric acid (HNO_3)	= 63.01	SG = 1.413

Flowchart for Testing Carbonates

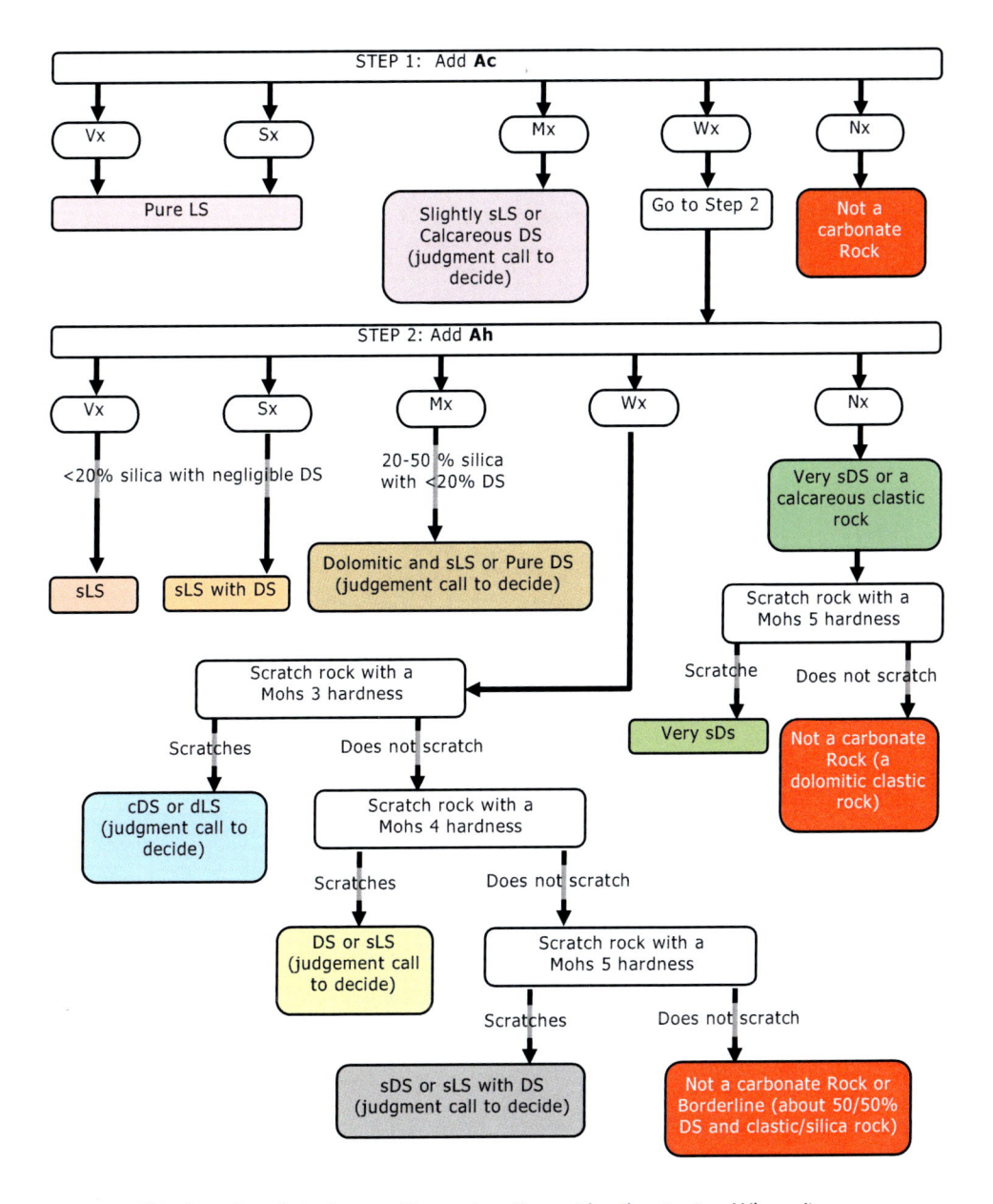

STEP 1: Add Ac

- **Vx** → Pure LS
- **Sx** → (leads to Pure LS area)
- **Mx** → Slightly sLS or Calcareous DS (judgment call to decide)
- **Wx** → Go to Step 2
- **Nx** → Not a carbonate Rock

STEP 2: Add Ah

- **Vx** (<20% silica with negligible DS) → sLS
- **Sx** → sLS with DS
- **Mx** (20-50% silica with <20% DS) → Dolomitic and sLS or Pure DS (judgement call to decide)
- **Wx** → Scratch rock with a Mohs 3 hardness
  - Scratches → cDS or dLS (judgment call to decide)
  - Does not scratch → Scratch rock with a Mohs 4 hardness
    - Scratches → DS or sLS (judgement call to decide)
    - Does not scratch → Scratch rock with a Mohs 5 hardness
      - Scratches → sDS or sLS with DS (judgment call to decide)
      - Does not scratch → Not a carbonate Rock or Borderline (about 50/50% DS and clastic/silica rock)
- **Nx** → Very sDS or a calcareous clastic rock
  - Scratch rock with a Mohs 5 hardness
    - Scratches → Very sDs
    - Does not scratch → Not a carbonate Rock (a dolomitic clastic rock)

NOTE: *This flowchart is to be used in conjunction with other tests. Where it says, "judgment call to decide", that does NOT mean other tests cannot be used. Microscopic analysis may bring to light non macroscopic features that you can use to definitely classify the rock one way or another.*

Crystalline Carbonates

Visually, limestone and dolostone look very similar under magnification. Both belong to the hexagonal (or trigonal, depending on who you ask) crystal systems. The only real difference is in their crystal habit. Both will form rhombohedral crystals, but only calcite will form "hexagonal scalenohedral" (or fat blades a.k.a. skinny upside down cones with angular sides). Although dolomite will form wide bladed crystals. Both minerals can be massive and both have white streaks. Dolomite is also harder with a Mohs of 3.5-4, as where calcite is a 3. Calcite defines the Mohs hardness of 3. It is common for Phanerozoic carbonates in the Midwest to be crystalline. Due to the angularity of most crystalline carbonates, we use a modified Wentworth scale.

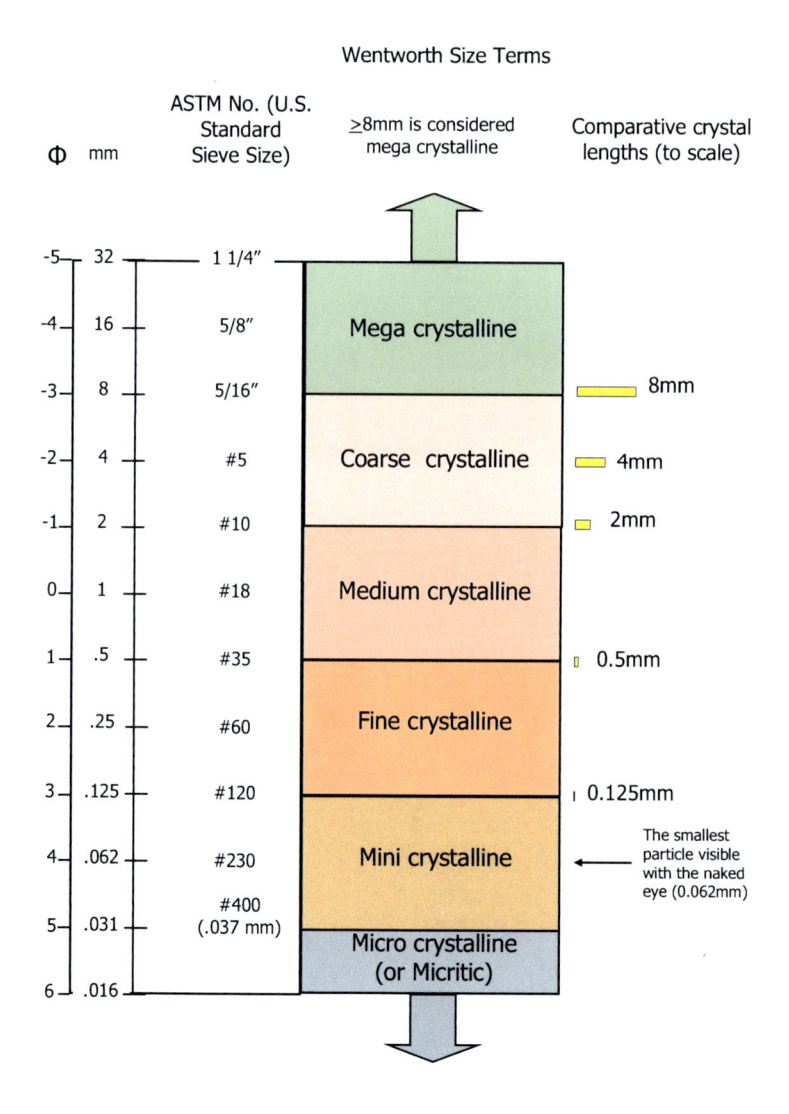

Wentworth Size Terms

Φ	mm	ASTM No. (U.S. Standard Sieve Size)	≥8mm is considered mega crystalline	Comparative crystal lengths (to scale)
-5	32	1 1/4"		
-4	16	5/8"	Mega crystalline	
-3	8	5/16"		8mm
-2	4	#5	Coarse crystalline	4mm
-1	2	#10		2mm
0	1	#18	Medium crystalline	
1	.5	#35		0.5mm
2	.25	#60	Fine crystalline	
3	.125	#120		0.125mm
4	.062	#230	Mini crystalline	The smallest particle visible with the naked eye (0.062mm)
5	.031	#400 (.037 mm)	Micro crystalline (or Micritic)	
6	.016			

Non Crystalline Grains

Fossils

The most common grains you are likely to see in a Phanerozoic carbonate other than crystals are fossils. If a carbonate has fossils they should be noted with words such as "trace fossils" <5%, "with fossils" 5-15%, or "fossiliferous" 15-35%. Dunham (1962) devised a classification scheme for fossil rich rocks and he treats them as grains (chart below). It addresses most of the Phanerozoic bio carbonates.

What it does not address are the Precambrian stromatolite rich carbonates. For those the crystalline carbonate classification can be used modified with the terms "trace stromatolites" <5% of the rock, "with stromatolites" 5-15% of the rock, "stromatolitic" 15-35% of the rock, and "stromatolite (insert appropriate carbonate name)", >35% of the rock.

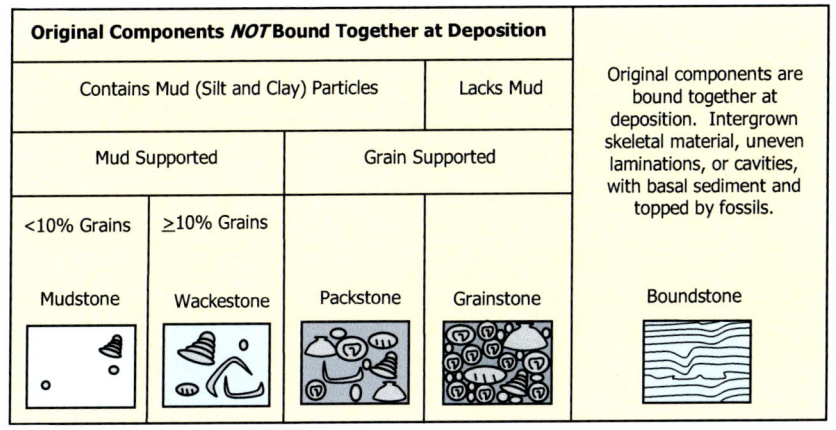

Original Components *NOT* Bound Together at Deposition				Original components are bound together at deposition. Intergrown skeletal material, uneven laminations, or cavities, with basal sediment and topped by fossils.
Contains Mud (Silt and Clay) Particles			Lacks Mud	
Mud Supported		Grain Supported		
<10% Grains	≥10% Grains			
Mudstone	Wackestone	Packstone	Grainstone	Boundstone

Adapted after Dunham (1962)

Clastics

Although rare, clastics can occur in carbonate rocks. You use "argillaceous" for silt and clay (a.k.a. fines). You use "sandy" for sand. A carbonate <5% clastics isn't mentioned. 5-35% clastics is "with (insert argillaceous or sandy)". 35-45% clastics is either sandy or argillaceous. If both sand and fines are present terms like "argillaceous micritic dolostone with sand", can be used. Clastics are almost always only going to be present in mini and micro crystalline carbonates. If you have ≥50% clastics, you don't have a carbonate rock. You have either a calcareous or dolomitic clastic rock.

Carbonate ternary plot based off mineralogy

(2nd order normalization)

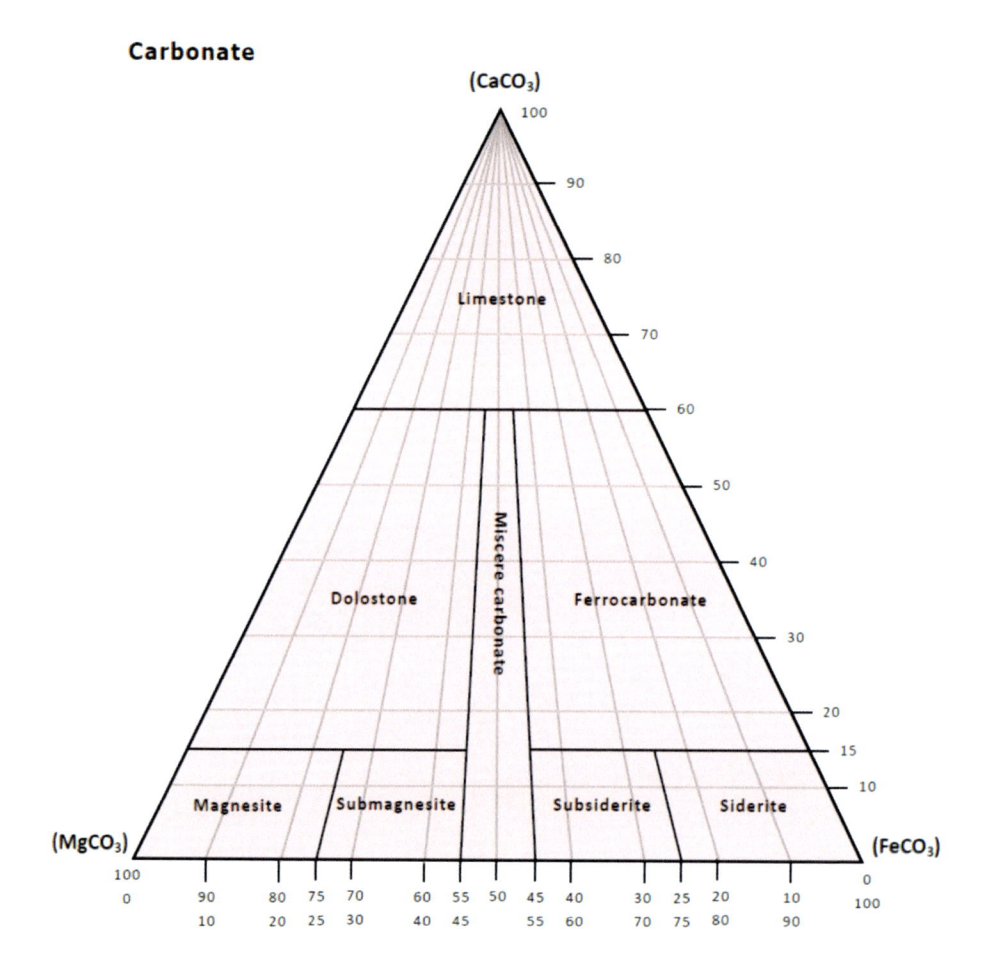

Carbonate
(CaCO$_3$)

Limestone

Dolostone

Miscere carbonate

Ferrocarbonate

Magnesite | Submagnesite | Subsiderite | Siderite

(MgCO$_3$)   (FeCO$_3$)

Miscere is Latin for mix

Lithological Carbonate Symbols

It is often easier to graphically depict a certain rock type in either field notes or a lab book. Below are lithological symbols for carbonates.

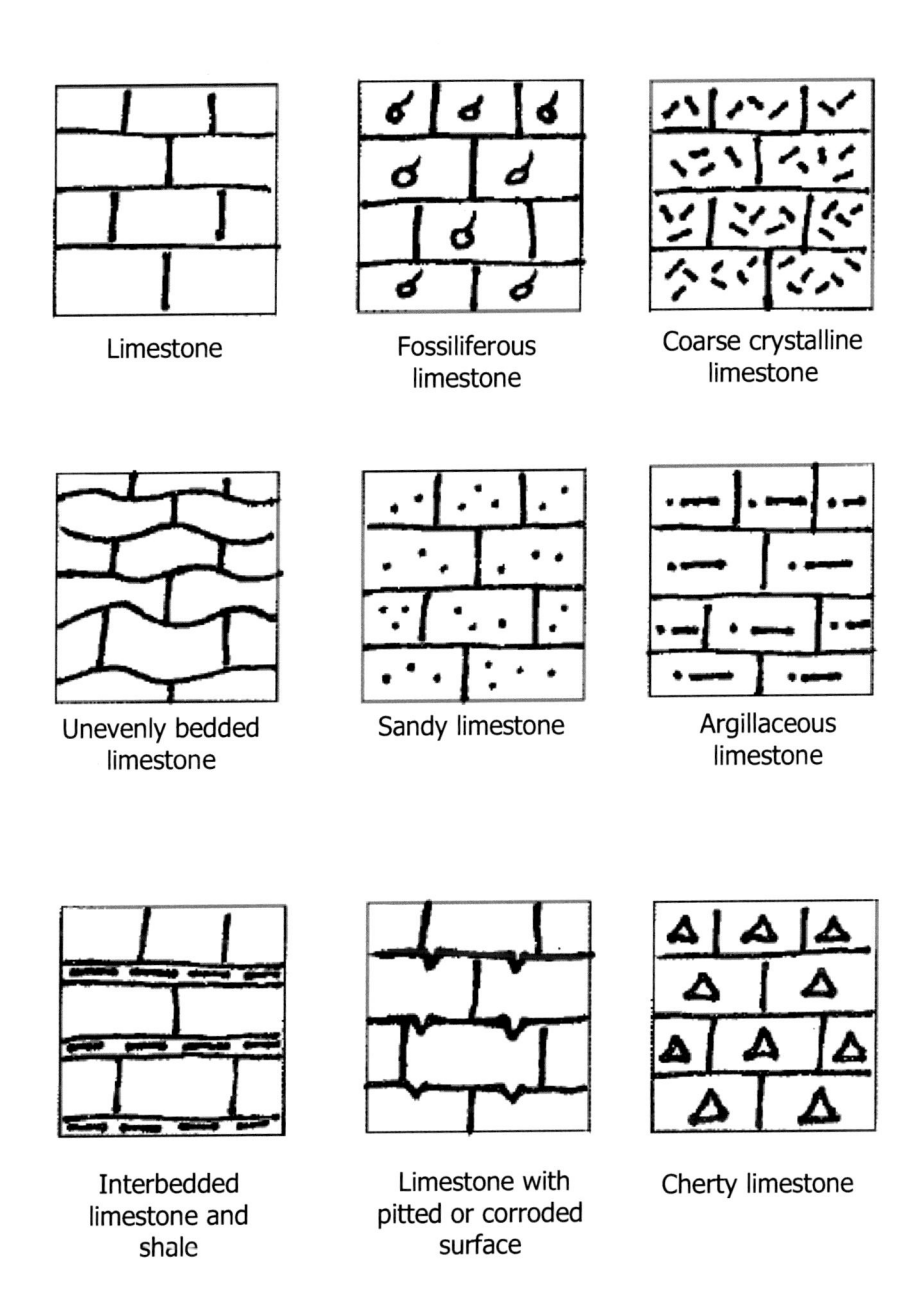

Limestone

Fossiliferous limestone

Coarse crystalline limestone

Unevenly bedded limestone

Sandy limestone

Argillaceous limestone

Interbedded limestone and shale

Limestone with pitted or corroded surface

Cherty limestone

Stromatolitic limestone

Nodular limestone

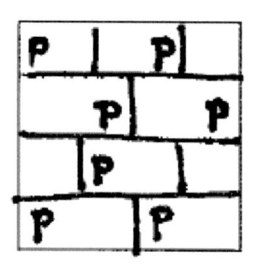

Pyritic limestone

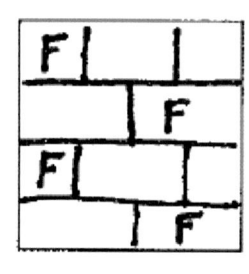

Ferrous limestone

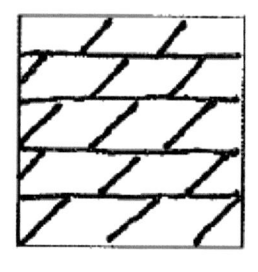

Dolostone (same variations as limestone)

CHERT

Chert is just micro crystalline siliceous rock. Like carbonates, it can be derived from biological processes (it will often contain fossils), chemical precipitation from solution, or mechanical weathering.

Phanerozoic Chert

In Phanerozoic rocks chert commonly exists as nodules, lenses, or thin beds; these three things are the occurrence within carbonate rock. When it is the lesser of the rock you would describe the occurrence, and slap either cherty or with chert to the name. e.g. "Nodular cherty dolostone" or a "Dolostone with lenticular chert". Of course you can always go above and beyond and give and give actual percentages. Rocks with >50% chert are called "chertstone". In this case, the minor carbonate becomes the modifier. e.g. "Thinly bedded dolomitic chertstone" or "chertstone with lenticular limestone".

Precambrian Chert

Chert in the Precambrian is a different monster. It almost is exclusively believed to be derived from wind blown loess (in the Paleoproterozoic glacial deposits) or as a chemical/biological deposit within banded iron formations. It does not commonly occur interbedded with carbonates as it does in the Phanerozoic carbonates, although it can. Regardless of its origin, it is usually massive, hard, micritic, and lacks any fossils. It's generally just called "chertstone" with any appropriate modifiers in a similar manner as carbonate and clastic rocks.

EVAPORITE

Evaporites include gypsum, anhydrite, and halite; these are common marine evaporites. Borax and epsomite are common lacustral ones. All known evaporite deposits in the Midwest are marine in origin. They occur only as thinner strata within carbonates and calcareous clastic rocks. There are no known Precambrian evaporites in the Midwest. In the Midwest, their presence is restricted to mainly Silurian though Permian deposits in the Forest City, Illinois, Michigan, and under Lake Huron off the fringes of the Michigan Basin (see colored areas below). Due to their relative scarcity, no detailed classification scheme will be given herein. They can be described in a similar manner as to how chert is described within Phanerozoic carbonates.

Midwest Basins

Basins are in pale orange

Adapted from: Baumann (2016), and Dean et al. (1989)

BANDED IRON FORMATION (BIF)

Banded iron formations (BIFs) are perhaps the most common chemical/biological rocks of the Precambrian. Here's something interesting about BIF. It is the only sedimentary rock Earth no longer produces! All BIFs are Precambrian (Lindsey, 2007).

Textbook BIF consist of interbedded dark gray hematite and red jasper (really just red chert), as in the pic below. BIF can be far more complex than that. It's also difficult to classify due to its interbedding nature. Herein BIF is defined as "*dominantly layered iron oxide minerals interbedded with clastics and/or carbonates*". I have been slowly developing a classification system over the years. The field version appears in my "Midwest Field Geologist's Reference Book (Deluxe Edition)" (2019). For lab identification a little more detail is needed. The classification system is still in progress, but the basic framework has been completed (next page).

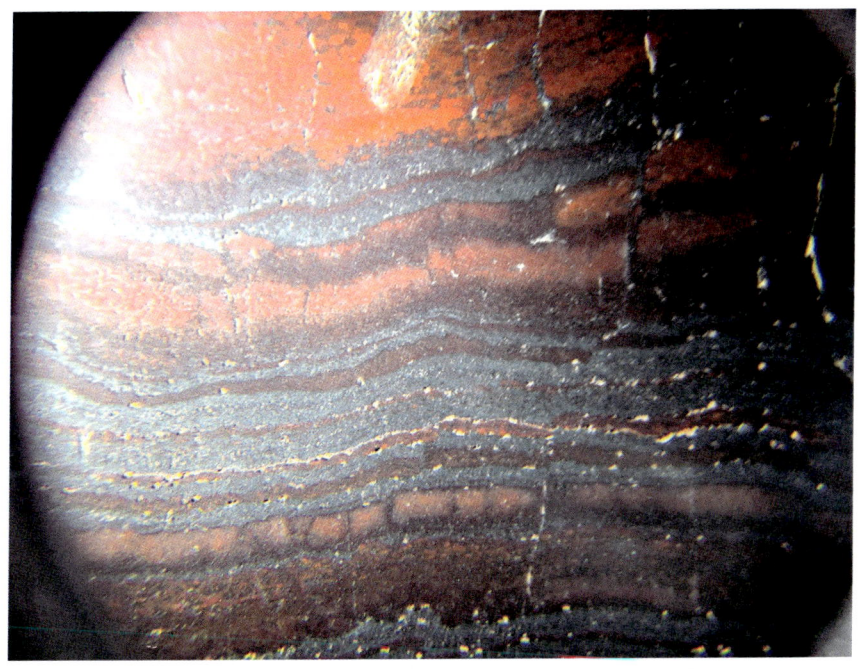

Semi polished hematite and red jasper BIF from the Upper Peninsula, Michigan.
40x magnification.

Silicate/Oxide/Carbonate Iron [SOC(Fe)] Diagram

The SOC(Fe) diagram plots the whole BIF rock for late diagenetic and low grade metamorphic BIF. This includes the interbedded parts. Is the rock more of an iron oxide (Fe-O) like hematite, goethite, or magnetite; an iron carbonate (Fe-C) like siderite, limestone, or dolostone; or is it silica (Fe-S) rich (quartz)? According to the Strunz classification (Mills, et al., 2009, and Strunz and Nickel, 2001), silica is technically an oxide. It can be a silicate depending on who you ask. Chemists like to call it an oxide. Geologists like to call it a tectosilicate. Primary quartz is often visible disseminated within the iron part of the rock or as a sand component. The silicates do not have to be crystalline, they can be micritic as well. This includes the silicate chert traditionally called jasper. Below is how you would plot the three on a traditional ternary diagram.

(2nd order normalization)

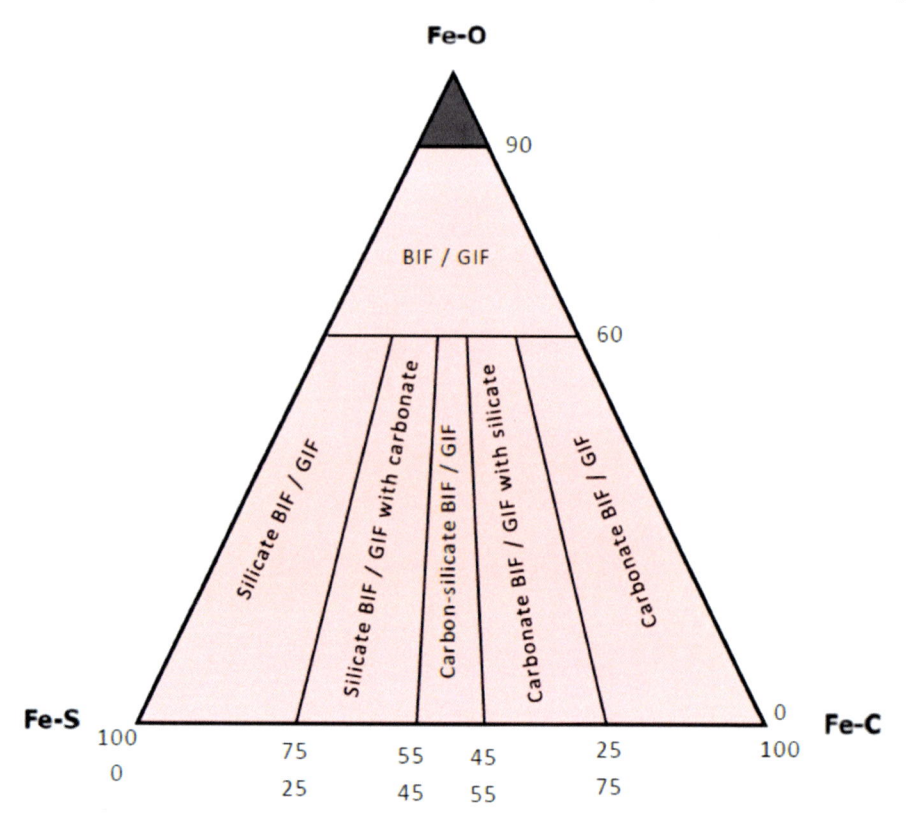

Fe-O

90

BIF / GIF

60

Silicate BIF / GIF
Silicate BIF / GIF with carbonate
Carbon-silicate BIF / GIF
Carbonate BIF / GIF with silicate
Carbonate BIF / GIF

Fe-S   100   75   55   45   25   0   **Fe-C**
             0   25   45   55   75   100

Use either BIF or GIF, which ever is appropriate. Don't use both.

Secondary Minerals

There are other minerals common in BIF that form either through weathering or diagenesis. Although they are not necessary to name the rock, you can use them to help identify the main part of the rock.

Ankerite Ca(Mg,Fe)(CO$_3$)$_2$ forms in the presence of hematite and dolomite. Its physical characteristics very closely resemble siderite and dolomite. In sedimentary rocks it forms through the low grade metamorphism of carbonate rich BIF.

Limonite FeO(OH)·nH$_2$O is a weathering product of iron rich rocks. It usually exists as a bright orange or yellow coating, almost powder like. It usually forms from the hydration of hematite and magnetite.

Magnesite MgCO$_3$ is a magnesium carbonate. It is very rare as a sedimentary rock, and rare in a meta-sedimentary rock. It reacts very weakly with cold dilute HCl but moderately with warm HCl. Although usually associated with ultramafic igneous rocks, it exists in BIF largely as a weathering product or replacement mineral of dolomite as the BIF undergoes low grade metamorphism.

Malachite Cu$_2$CO$_3$(OH)$_2$ is a weathering product of copper ore within a usually carbonate rich BIF. Copper isn't usually associated with BIF, but can be introduced and disseminated within through later hydrothermal activity. Especially in the Lake Superior area, where copper deposits are common. Its presence on BIF indicates the presence of disseminated copper within the BIF, as well as the presence of other carbonates. Azurite Cu$_3$(CO$_3$)$_2$(OH)$_2$ is a very blue mineral related to malachite that can theoretically also be present. However, azurite is not stable in open air and would quickly weather to malachite by reacting with the atmosphere.

Pyrite FeS$_2$ can form from the weathering of BIF if sulfides are introduced. Although rare with BIF, its presence may also indicate the presence carbonates like limestone; since pyrite, limestone, and dolostone commonly occur together in Phanerozoic sedimentary rocks.

Other BIF Components

Modifiers can be used such as "argillaceous, silty, sandy, conglomeratic, etc.", where appropriate.

When jasper is present as the silicate (usually as micro crystalline quartz or chalcedony), it should be described as a chert, since jasper is not really a rock type or mineral. The term jasperlite should be totally avoided as it is a gem term. Although the term jasper is usually used to describe red chert, the chert in BIF is commonly purple, brown, and even gray.

COAL

Coal is a rock deposited directly by dead plant material, that usually burns when a flame is applied.

Types of Coal

The classification of coal can potentially get complex, so a modified mudstone chart will be used. This classification deals only with bituminous coal. It is the only coal you will run into in the Midwest. There are no known metamorphic coals (anthracite) in the Midwest. In the Midwest, all coal is Carboniferous in age and contains a lot of sulfur. Identifiable plant and animal fossils are also common.

Coal is usually laminated and is a mixture of organic debris and mud that cannot be easily differentiated. It is formed from the compaction and decay of peat. In this respect it is unique and a wholly organic rock. Although coal is a rock, it the organic component is not made of minerals, as is the case of almost all other rock. In geology most minerals are inorganic by definition. If a rock is an aggregate of one or more minerals, why is coal considered a rock? This is a hold over from before we had developed crystallography and even really before we could chemically identify things. Coal has been known to occur within other rock since ancient times, so it has always been considered a rock. It is also subject to the rock cycle as are all other rocks.

Name	Definition
COAL	>50% of the grains are clay to very fine silt sized
SANDY COAL	35% to 50% of the grains are very fine to very coarse sand sized, with almost no silt
SILTY COAL	35% to 50% of the grains are very fine to coarse silt sized, with almost no sand
SANDY COAL with SILT	35% to 50% of the grains are very fine to very coarse sand sized, with 5% to <35% silt
SILTY COAL with SAND	35% to 50% of the grains are very fine to coarse silt sized, with 5% to <35% sand
CONGLOMERATIC COAL	35% to 50% of the grains are granular sized or larger, the rest is sand or finer
COAL with CONGLOMERATE	5% to <35% of the grains are granular sized or larger, the rest is sand or finer

NOTE: *You may be wondering why some of these percentages add up to more than the first listed 50% requirement. It's due to the variability and relative ratios of the constituents that are <50%.*

Other Coal Components

Sulfur is a major component of coal in the Midwest. There are places in southern Illinois where it weathers out bright yellow and the air smells like rotten eggs on a hot and still summer day.

There are relatively simple ways to determine this in a chemistry lab. However, those techniques still require special equipment (Fuerstenau, 1994). Sulfur can be either primary or secondary and come in many different forms; "total sulfur (organic and inorganic sulfur in coal and carbonaceous products), ash sulfur (sulfur present in ash of fuels), flammable sulfur (the difference between total sulfur and ash sulfur), sulfate sulfur (gypsum, epsomite, jarosite, melanteritet), pyrite sulfur (as pyrite and/or markasite), inorganic sulfur (sulfates and sulfides), organic sulfur (part of the organic matter of the coal, the difference between total and inorganic sulfur), and elemental sulfur (occurs as chemical elements S)" (Mahapatra, 2016).

Visually, there's really only two easy sulfur tests. You can look for pyrite under magnification or you can burn it to determine flammable sulfur. If you choose to burn it use your nose to determine the appropriate sulfur modifier. If there is no rotten egg smell then you just don't mention sulfur.

Modifier and type	Definition
Slightly sulfuric COAL	Hint faint of a rotten egg smell
Sulfuric COAL	Rotten egg smell is easily noticeable
Heavily sulfuric COAL	Rotten egg smell is strong
Sulfur COAL	Smell of rotten eggs is potent

NOTES: *Sulfur coal will often have visible yellow sulfur that contrasts sharply with the black coal in between laminations. If warm enough out, it can even be smelled without burning it.*

You may be wondering why some of these percentages add up to more than the first listed 50% requirement. It's due to the variability and relative ratios of the constituents that are <50%.

Lithological Chertstone, BIF, Evaporite, and Coal Symbols

It is often easier to graphically depict a certain rock type in either field notes or a lab book. Below are lithological symbols for the rest of the non clastic rocks.

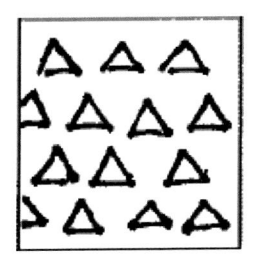

Chertstone

Evaporite

Banded Iron Formation

Coal

Sedimentary Rocks Under High Magnification:

This is a sandstone cobble I found in a creek. Although I can't identify its formation, I can work out the mineralogy. The diameter of the circle is ~3mm. The dark and bluish colors are iron staining and the mineralogy is mostly feldspar, as are the red grains. The only quartz are the grains without internal black specks, and they make up only about 15% of the rock. There are very little lithics and practically no matrix. The sandstone is a poorly sorted arkosic arenite.

This is a sandstone was a loose block in Houghton Michigan. It is a piece of the Jacobsville Group. The diameter of the circle is ~3mm. The red color is mostly primary iron staining. The small black specks are a mix of mafic and iron lithics. This sample is better sorted and contains more quartz (~1/3 of the rock) than the above sample. The yellows and more orange grains as well as the opaque white grains are feldspars. It's a lithic arkosic arenite.

Sedimentary Rocks Under High Magnification:

This is Cambrian aged "greensand" from West Salem Wisconsin, is rich in glauconite. The green is NOT olivine. It's technically a lithic arenite because glauconite counts as lithic grains. It is poorly to moderately sorted and there is a gap in grading.

Diameter ~3 mm

The green color is still visible under moderate magnification. Most of the grains are glauconite. Most of the rest of the sand is quartz with minor lithics of mostly limestone fragments. There is less than 10% feldspar.

Diameter ~0.9 mm

At high magnification under visible light, the glauconite loses its green color and appears more gray.

Sedimentary Rocks Under High Magnification:

This is BIF from the Upper Peninsula of Michigan. The diameter of the circle is ~3.5 mm. Deep red is often called jasper, but its just non crystalline to micro crystalline red chert a.k.a. chalcedony. The gray part is specular hematite. The pale yellow to almost white veins are non crystalline SiO_2.

This is ejecta from the Sudbury impact found outside of Thunder Bay Ontario. The diameter of the circle is ~3mm. Most grains are subangular to rounded and set in a gray carbonate to clastic matrix. Shocked quartz is present in the smaller grains. Quartz varies greatly in color from translucent white to deep red, and is only present in the smaller grains. However, most of the deep red and brown smaller grains are not quartz. The nearly black fine grains are mafic volcanics. The larger granular to fine pebbles tend to be carbonate and siliceous mud; and other lithic fragments such as sandstones and felsic volcanics.

This page is intentionally left blank

This page is intentionally left blank

Lab Reference Book

Tables and charts applicable to all rock types or general information

Previous page is a Geode with calcite and dolomite crystals from southwestern Indiana

Introduction

There are many tests and methods that apply to rocks and minerals regardless of rock type. Herein is included some of those helpful things. Such concepts as color, density, crystal systems, particle volume, and optical techniques. Knowing the interior of the Earth is also helpful for interpretations. A section was included on Earth's interior.

Earth's Interior

Geothermal Gradient

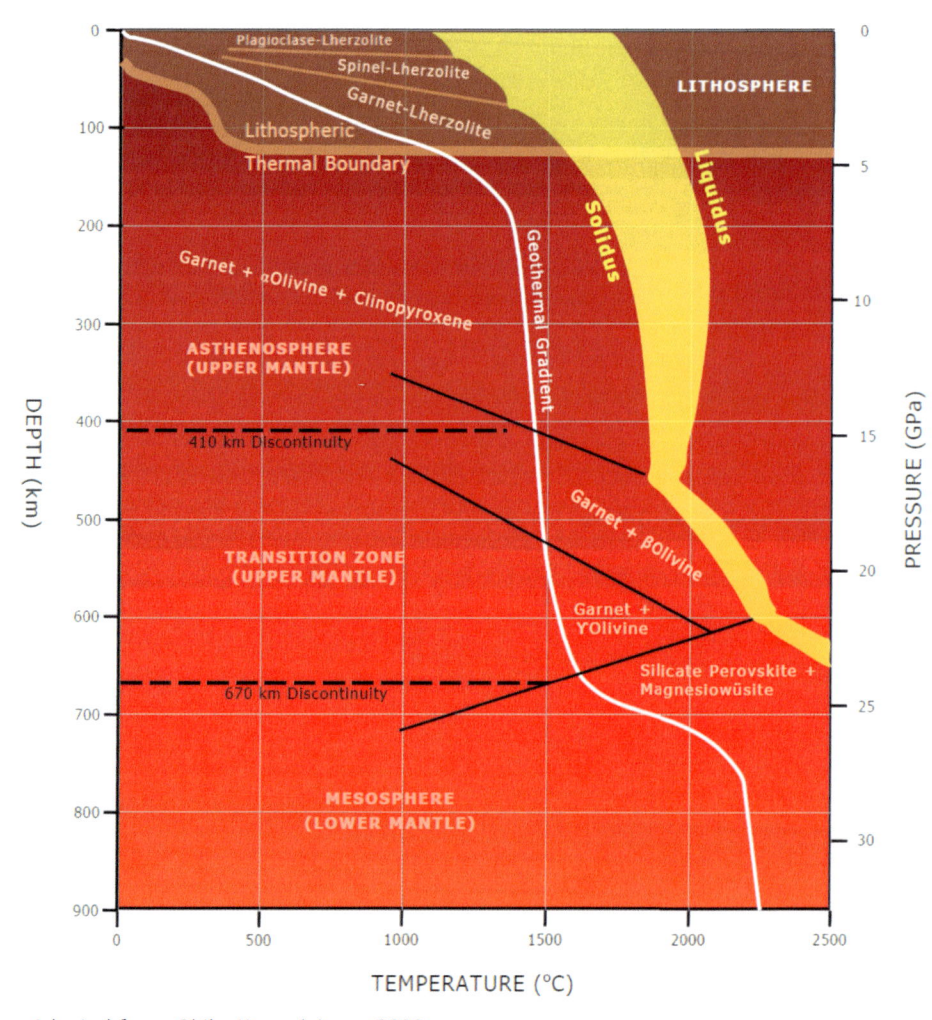

This is the geothermal gradient into the earth represented by a cross section from sea level (0 km) to a depth of 900 km. This model assumes at least a partially 2 layer mantle convection. The two continuities are placed at slightly different levels, depending on your source. It is likely that the levels are not exactly at the same depth, just like the lithosphere s not. 410 discontinuity is likely from 350 to 410 km. The 670 discontinuity varies from 660 to 700 km.

Internal Mechanical Layers of the Earth and Their Percent Mass

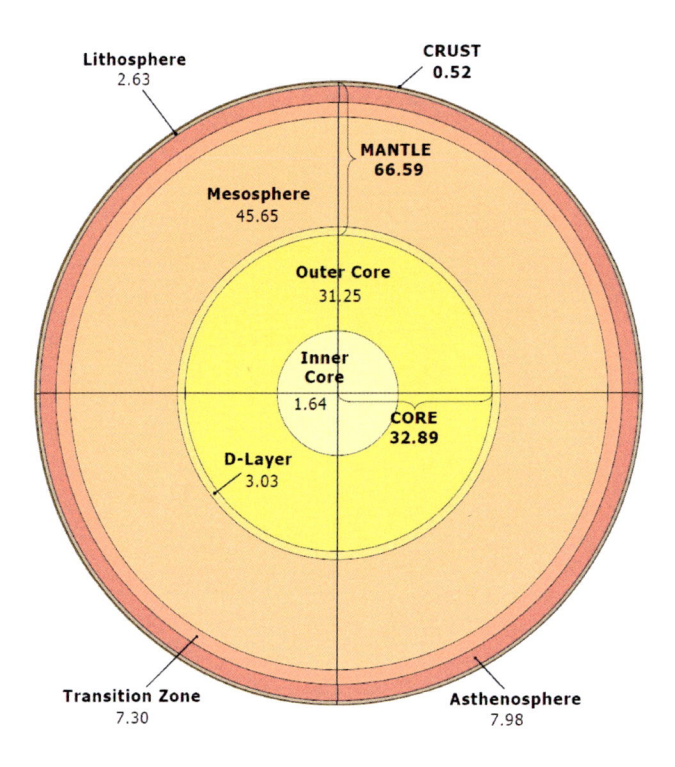

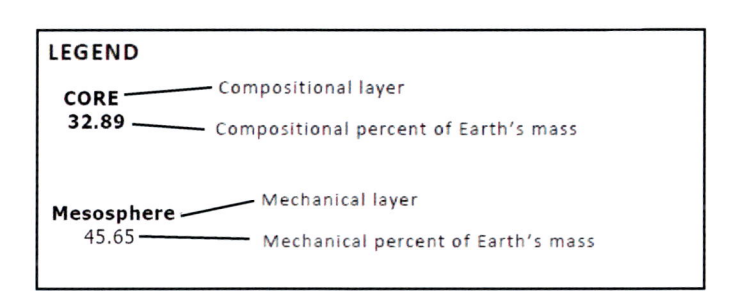

Lithosphere 2.63
CRUST 0.52
Mesosphere 45.65
MANTLE 66.59
Outer Core 31.25
Inner Core 1.64
CORE 32.89
D-Layer 3.03
Transition Zone 7.30
Asthenosphere 7.98

LEGEND

CORE — Compositional layer
32.89 — Compositional percent of Earth's mass

Mesosphere — Mechanical layer
45.65 — Mechanical percent of Earth's mass

I calculated these numbers back in 2018 and it worked out very well. These percents were used in conjunction with densities and the mass calculations are near a dead match for NASA's mass of the Earth. Those calculations summaries are on p241-243.

Internal Mechanical Layers of the Earth and Their Percent Volume

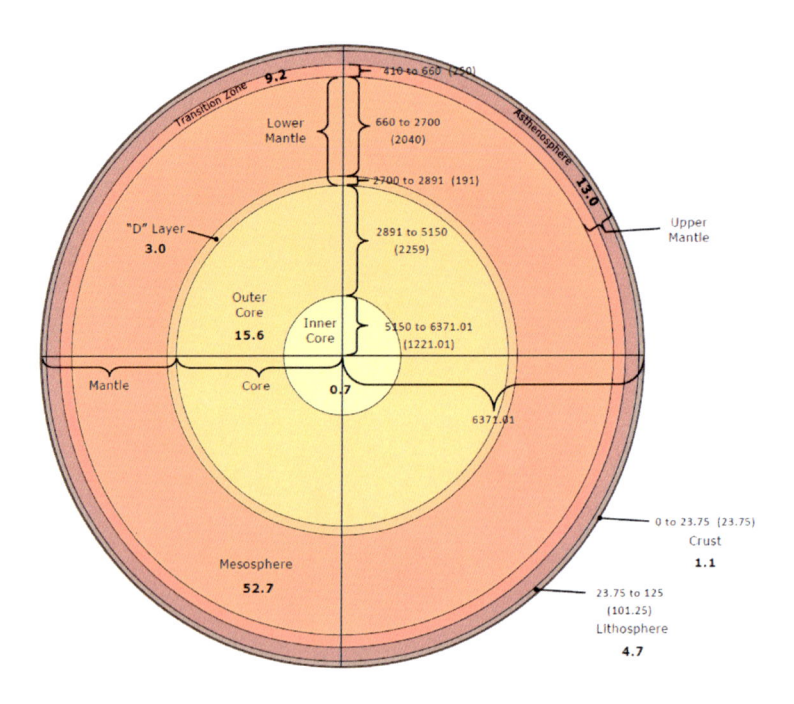

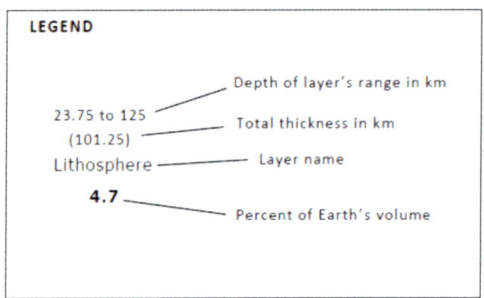

Transition Zone **9.2**
410 to 660 (250)
Lower Mantle
660 to 2700 (2040)
Asthenosphere **13.0**
2700 to 2891 (191)
Upper Mantle
"D" Layer **3.0**
2891 to 5150 (2259)
Outer Core **15.6**
Inner Core
5150 to 6371.01 (1221.01)
Mantle
Core
0.7
6371.01
0 to 23.75 (23.75)
Crust **1.1**
Mesosphere **52.7**
23.75 to 125 (101.25)
Lithosphere **4.7**

LEGEND

23.75 to 125 ——— Depth of layer's range in km
(101.25) ——— Total thickness in km
Lithosphere ——— Layer name
4.7 ——— Percent of Earth's volume

A more realistic cross section of the interior of the Earth

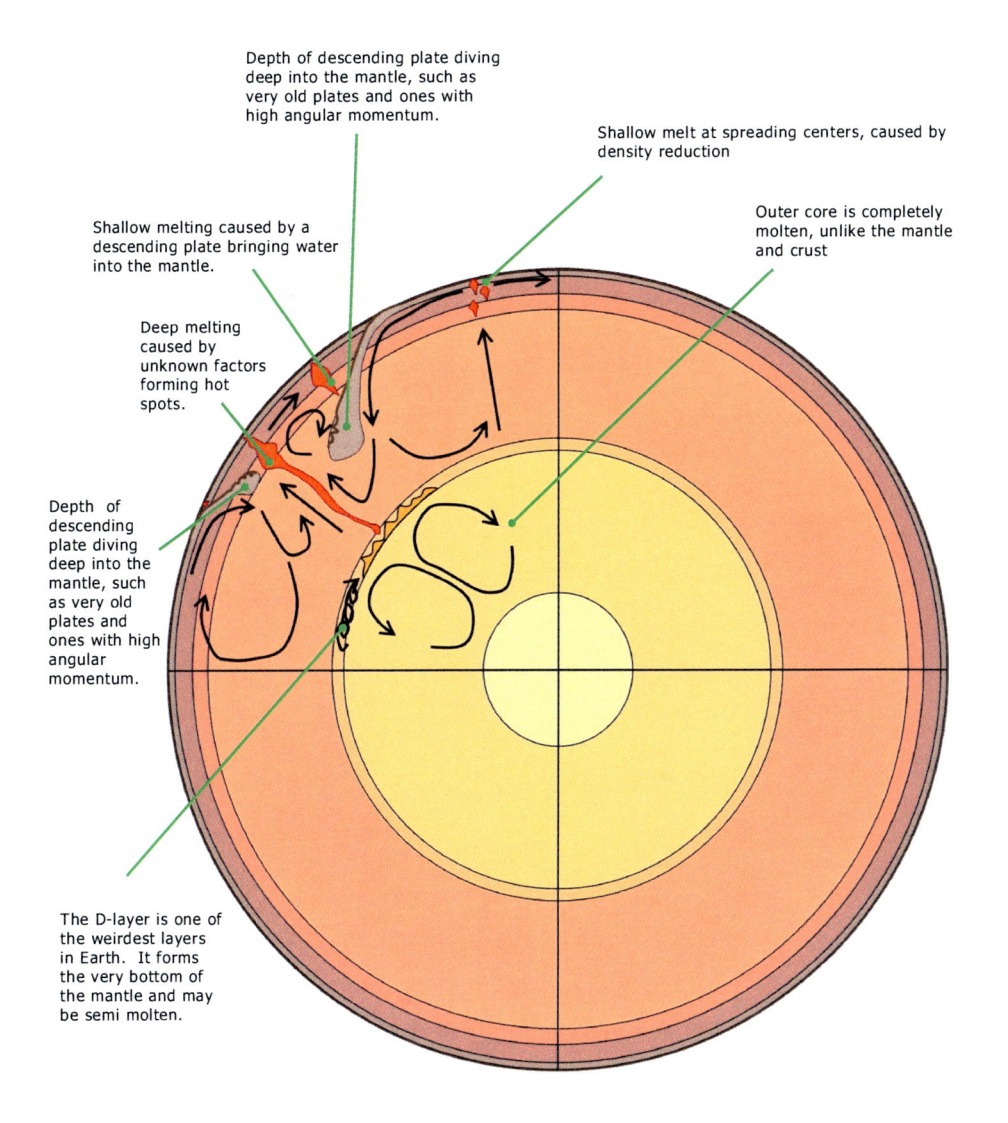

Depth of descending plate diving deep into the mantle, such as very old plates and ones with high angular momentum.

Shallow melt at spreading centers, caused by density reduction

Shallow melting caused by a descending plate bringing water into the mantle.

Outer core is completely molten, unlike the mantle and crust

Deep melting caused by unknown factors forming hot spots.

Depth of descending plate diving deep into the mantle, such as very old plates and ones with high angular momentum.

The D-layer is one of the weirdest layers in Earth. It forms the very bottom of the mantle and may be semi molten.

Layering colors correspond to those on p.239-240

Hypothetical convective cells

Basic Mechanical Structures in Earth's Interior

Crust (CR)
Outer "shell" of the Earth consisting of continental and ocean basin solid rock. Rigid layer involved directly in plate tectonics, especially subduction.

Lithosphere (LI)
Upper most mantle. Also a solid. Dominantly olivine. Rigid layer involved directly in plate tectonics, especially subduction. Composed of metamorphic crustal rocks and olivine.

Asthenosphere (AS)
Solid upper mantle that is involved in plate tectonic movement and isostatic adjustments. Likely highly faulted and may be a source of shallow mantle plumes. Mostly olivine, pyroxene, and garnet.

Transitional Zone (TZ)
Solid ductile upper mantle that is convective due to subducting plates. Mostly spinels and majorite. Likely the lower most depth where hydrate minerals are found.

Mesosphere (ME)
Solid ductile to rigid lower mantle that is the dying place for most lithospheric plates. Mostly perovskite and ferropericlase. The foci of the deepest earthquakes occur at the top of this layer. Origin of deep magma plumes.

"D" Layer (DL)
A very viscous layer at the core-mantle boundary. Mostly solid to partially molten. Its internal workings and composition are not known for sure. It is likely post-perovskite and ferropericlase. It shows great topographic relief.

Outer Core (OC)
A molten and highly convective layer. It is involved in generating Earth's magnetic field. It is likely composed of mostly iron and nickel, with some sulfides.

Inner Core (IC)
A solid and rigid layer, extending to the center of the planet. It is involved in generating Earth's magnetic field. It is likely composed of mostly iron and nickel, with some sulfides.

Using density to calculate mass

We need to convert out percentages into decimals

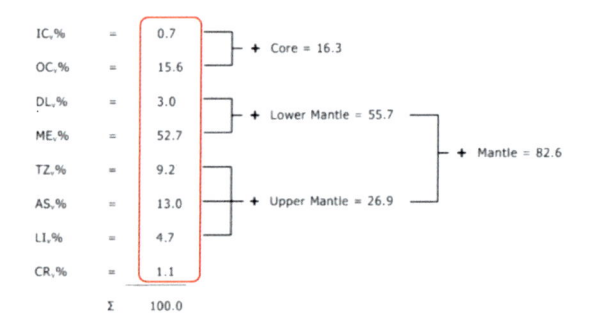

IC,%	=	0.7
OC,%	=	15.6
DL,%	=	3.0
ME,%	=	52.7
TZ,%	=	9.2
AS,%	=	13.0
LI,%	=	4.7
CR,%	=	1.1
Σ		100.0

- Core = 16.3 (IC + OC)
- Lower Mantle = 55.7 (DL + ME)
- Upper Mantle = 26.9 (TZ + AS + LI + CR)
- Mantle = 82.6

Using density to calculate mass

We need to convert out percentages into decimals

IC,%	=	0.7
OC,%	=	15.6
DL,%	=	3.0
ME,%	=	52.7
TZ,%	=	9.2
AS,%	=	13.0
LI,%	=	4.7
CR,%	=	1.1
Σ		100.0

- Core = 16.3
- Lower Mantle = 55.7
- Upper Mantle = 26.9
- Mantle = 82.6

$\div 100 =$

- 0.007 (IC_v)
- 0.156 (OC_v)
- 0.030 (DL_v)
- 0.527 (ME_v)
- 0.092 (TZ_v)
- 0.130 (AS_v)
- 0.047 (LI_v)
- 0.011 (CR_v)

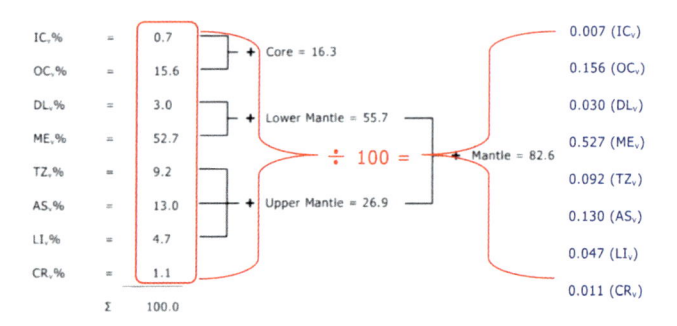

Using density to calculate mass of Earth

Now that we have our volume ratio, we need the average mean density of each layer.

	Density (g/cm³)			Average Density (g/cm³)		Average Density (g/cm³) x V
0.007 (IC_v)	12.8 to 13.1			13.0	(IC_p)	0.091 (IC_{p1})
0.156 (OC_v)	9.9 to 12.2			11.1	(OC_p)	1.7316 (OC_{p1})
0.030 (DL_v)	5.6	⎤	Range of the	5.6	(DL_p)	0.168 (DL_{p1})
0.527 (ME_v)	4.4	⎦	lower mantle	4.8	(ME_p)	2.5296 (ME_{p1})
0.092 (TZ_v)	4.4	⎤	Range of the	4.4	(TZ_p)	0.4048 (TZ_{p1})
0.130 (AS_v)	3.4	⎦	upper mantle	3.4	(AS_p)	0.442 (AS_{p1})
0.047 (LI_v)	2.7 to 3.4			3.1	(LI_p)	0.1457 (LI_{p1})
0.011 (CR_v)	2.2 to 2.9			2.6	(CR_p)	0.0286 (CR_{p1})

$\Sigma = 5.5413$ g/cm³ (ave.)

How does my average density compare with the value given by NASA?

5.5413 g/cm³ (ave.) = my calculated value

5.514 g/cm³ = value given by NASA https://nssdc.gsfc.nasa.gov/planetary/factsheet/earthfact.html

Average Density (g/cm³) x V / [g/cm³ (ave.)]		($p1/ap$) x $5.9723 \cdot 10^{24}$kg
0.091 / 5.5413 = 0.0164	($IC_{p1/ap}$)	(IC_m) = 0.097946 = m($\cdot 10^{24}$kg) = $9.7946 \cdot 10^{22}$kg
1.7316 / 5.5413 = 0.3125	($OC_{p1/ap}$)	(OC_m) = 1.866344 = m($\cdot 10^{24}$kg) = $1.8663 \cdot 10^{24}$kg
0.168 / 5.5413 = 0.0303	($DL_{p1/ap}$)	(DL_m) = 0.180961 = m($\cdot 10^{24}$kg) = $1.8010 \cdot 10^{23}$kg
2.5296 / 5.5413 = 0.4565	($ME_{p1/ap}$)	(ME_m) = 2.726355 = m($\cdot 10^{24}$kg) = $2.7264 \cdot 10^{24}$kg
0.4048 / 5.5413 = 0.0730	($TZ_{p1/ap}$)	(TZ_m) = 0.435978 = m($\cdot 10^{24}$kg) = $4.3598 \cdot 10^{23}$kg
0.442 / 5.5413 = 0.0798	($AS_{p1/ap}$)	(AS_m) = 0.47659 = m($\cdot 10^{24}$kg) = $4.7659 \cdot 10^{23}$kg
0.1457 / 5.5413 = 0.0263	($LI_{p1/ap}$)	(LI_m) = 0.157071 = m($\cdot 10^{24}$kg) = $1.5707 \cdot 10^{23}$kg
0.0286 / 5.5413 = 0.0052	($CR_{p1/ap}$)	(CR_m) = 0.031056 = m($\cdot 10^{24}$kg) = $3.1056 \cdot 10^{22}$kg

$\Sigma = 1.0000$ = total mass

$\Sigma = 5.972369 \cdot 10^{24}$kg ~$5.9724 \cdot 10^{24}$kg; NASA's = $5.9723 \cdot 10^{24}$kg

Layers of the Earth in Percent Volume

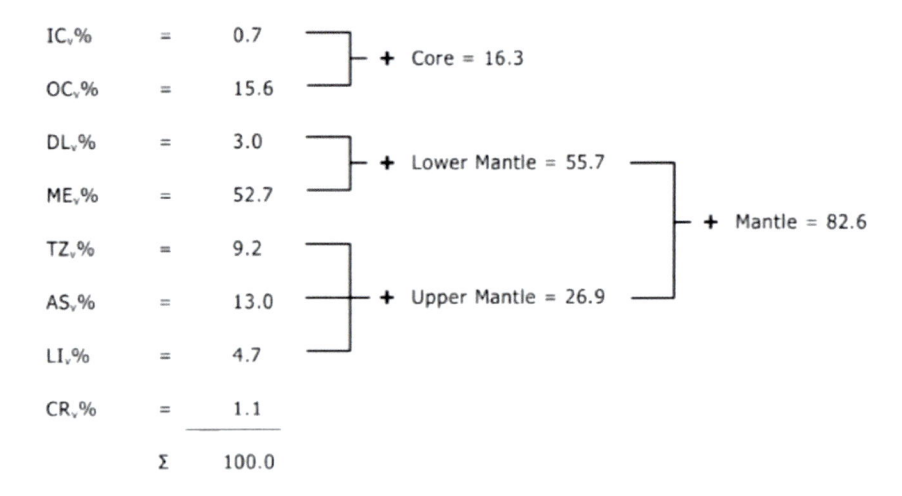

$IC_v\%$	=	0.7	⎤
			⎦ + Core = 16.3
$OC_v\%$	=	15.6	⎦
$DL_v\%$	=	3.0	⎤
			⎦ + Lower Mantle = 55.7 ⎤
$ME_v\%$	=	52.7	⎦ ⎥
			⎥ + Mantle = 82.6
$TZ_v\%$	=	9.2	⎤ ⎥
			⎥
$AS_v\%$	=	13.0	⎥ + Upper Mantle = 26.9 ⎦
			⎥
$LI_v\%$	=	4.7	⎦
$CR_v\%$	=	1.1	
	Σ	100.0	

Earth Fact Sheet

Bulk parameters

Mass (10^{24} kg)	5.9723
Volume (10^{10} km^3)	108.321
Equatorial radius (km)	6378.137
Polar radius (km)	6356.752
Volumetric mean radius (km)	6371.008
Core radius (km)	3485
Ellipticity (Flattening)	0.00335
Mean density (kg/m^3)	5514
Surface gravity (m/s^2)	9.798
Surface acceleration (m/s^2)	9.780
Escape velocity (km/s)	11.186
GM (x 10^6 km^3/s^2)	0.39860
Bond albedo	0.306
Geometric albedo	0.434
Black-body temperature (K)	254.0
Number of natural satellites	1
Planetary ring system	No

Orbital parameters

Semimajor axis (10^6 km)	149.60
Sidereal orbit period (days)	365.256
Tropical orbit period (days)	365.242
Perihelion (10^6 km)	147.09
Aphelion (10^6 km)	152.10
Mean orbital velocity (km/s)	29.78
Max. orbital velocity (km/s)	30.29
Min. orbital velocity (km/s)	29.29
Orbit inclination (deg)	0.000
Orbit eccentricity	0.0167
Sidereal rotation period (hrs)	23.9345
Length of day (hrs)	24.0000
Obliquity to orbit (deg)	23.44
Inclination of equator (deg)	23.44

https://nssdc.gsfc.nasa.gov/planetary/factsheet/earthfact.html

ACCEPTABLE COLOR DESCRIPTIONS
Color of Lithological Units

Color is somewhat subjective and color ranges can be used. Since most of us do not carry Munsell charts with us, because they are expensive, we rely on our eyes. A soil's color should be described at its natural moisture content. A rock color should be based off the color of the rock when dry to the touch.

There are acceptable colors to use and unacceptable colors to use. You should stick with colors on the Munsell chart. Unaccepted colors people commonly use are "tan", which is usually a "light yellow brown". "Buff", which is usually a very light to nearly white yellow brown or a very pale brown. "Chocolate brown", which is usually brown or dark brown. "Rusty", which is almost always brown red or a dark red brown. "Midnight black" is just black, anything other than this is a shade of gray. Other bizarre colors I have seen but cannot translate are "fuchsia, canary, teal, and cardinal". Things like tan, buff, and canary are completely subjective and should not be used as colors or color modifiers. Tan is something you get at the beach. Buff is what you do to polish your car, or to describe a muscular person. Canary is a bird and the yellow ones have various color ranges.

Modifiers are acceptable and can be used in combination with other modifiers. Pale, light, dark, and very can be used with other modifiers. For example, very dark or very light. Where colors are splotchy and variable with distinct borders the term "mottled" can be used. For example, light gray mottled orange brown. The dominant color is always mentioned first. Suffixes like "-ish" are left out. "Pale" means nearly white. "Very light" is actually darker than "pale" and "light" is darker than "very light".

Just the color itself is used for mid-tones and no modifiers are needed. For example, "medium gray" or "moderate orange" are *NOT* valid. You would just say "gray" or "orange".

Gradations and variations with no distinct borders can also be used. For example, brown to red purple.

Below is a chart with acceptable modifiers and colors. The Munsell color chart does recognize the term "olive". I generally avoid it. However, I have used "olive brown" or "olive gray" in the past. I also will avoid the term pink in favor of light red. It is also acceptable for colors that are not quite white or black (but close) to use the term "nearly". For example, very light yellow to nearly white or very dark brown to nearly black.

For the purpose of identifying rocks, black and white are considered colors.

Acceptable Modifiers
Very
Light
Dark
Pale
Bright

Acceptable Modifiers for rocks with a metallic luster
Drab
Metallic

Drab and metallic are to be used only for rocks with a metallic luster. In the Midwest, this is mostly restricted to banded iron formations, native copper deposits, and highly metamorphic rocks.

Color name	RGB index	Visual color
Pink	255 140 165	
Red	255 0 0	
Orange	255 128 0	
Yellow	255 255 0	
Green	0 255 0	
Blue	0 0 255	
Purple	155 0 255	
Brown	150 75 0	
Olive	150 185 110	
Black	0 0 0	
White	255 255 255	
Gray	128 128 128	

Common colors of rocks

- Red brown
- Orange brown
- Light pink
- Pale red
- Gray red
- Pale orange
- Light Brown
- Brown
- Dark Brown
- Light yellow brown
- Gray brown
- Brown gray
- Orange yellow
- Yellow orange
- Yellow brown
- Olive
- Yellow olive
- Brown olive
- Pale olive
- Olive gray
- Pale yellow

Very light yellow	Light yellow	Dark yellow
Yellow green	Pale green	Gray green
Dark green	Very dark green	Pale green blue
Pale blue	Light gray blue	Blue gray
Gray blue	Dark gray blue	Purple blue
Pale purple	Dark purple	Pale purple red
Light purple red	Gray purple red	White to very light gray
Very light gray	Light gray	Gray
Dark gray	Very dark gray	Black

GRAIN PERCENTAGES BASED ON VOLUME

Most igneous rocks cannot be mechanically broken down to run a comparative analysis using weight or mass. So instead we use volume based off area within our lens. The following charts represent how you would see dark and/or light minerals under a round hand lens or under the microscope.

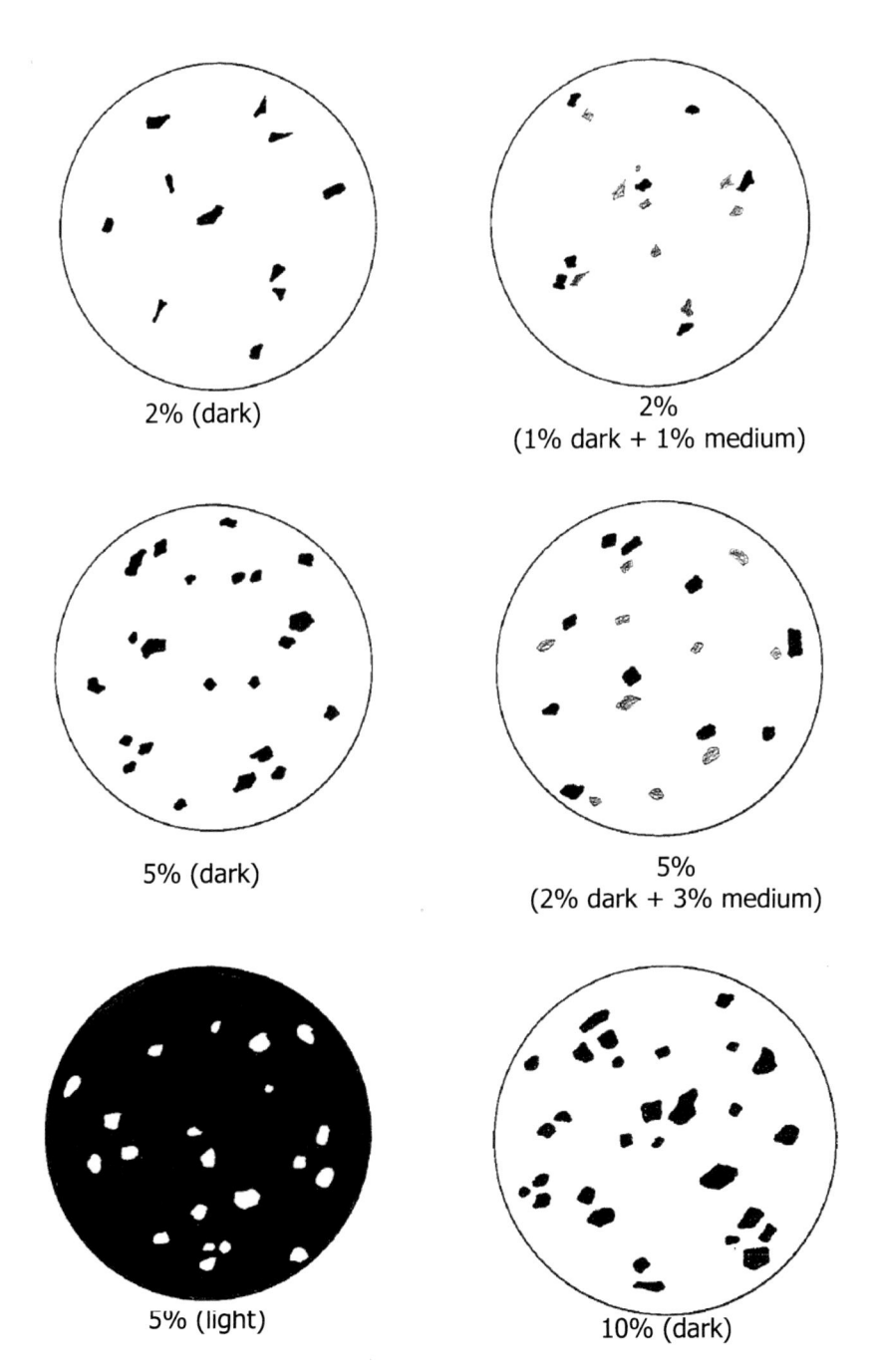

2% (dark)

2%
(1% dark + 1% medium)

5% (dark)

5%
(2% dark + 3% medium)

5% (light)

10% (dark)

248

10%
(5% dark + 5% medium)

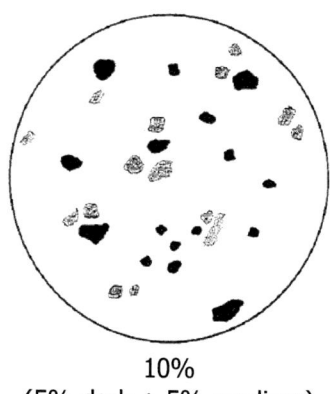

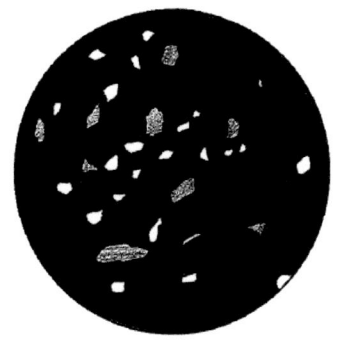

10%
(5% white + 5% medium)

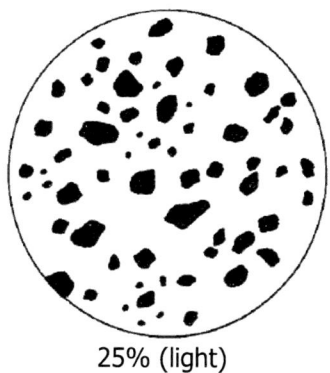

25% (light)

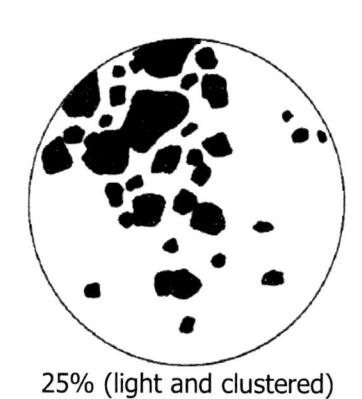

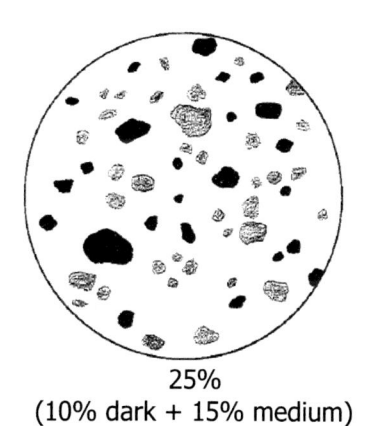

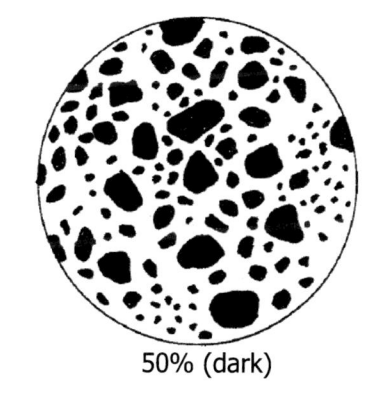

25% (light and clustered)

25%
(10% dark + 15% medium)

50% (dark)

50% (light)

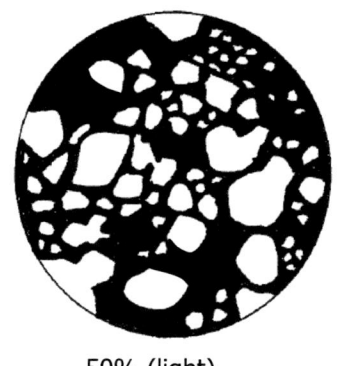

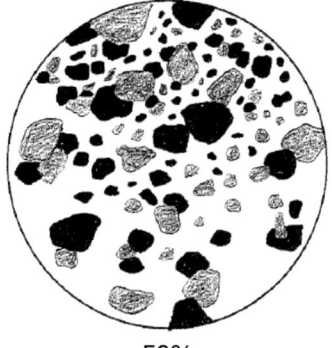

50%
(25% dark + 25% medium)

This takes a lot of practice. Once you are good at it and take your time, you can get percentages to within ±1.5%. That depends on many factors like magnification, the size of the area you're looking at, and the amount of time you have to look. Just remember, you are *NOT* counting grains here. You are estimating the percentage of area taken up by the grains you are studying.

This page is intentionally left blank

Basic properties of select rocks and minerals/mineraloids

Mineral or rock "/" indicates alternate name for the same mineral or rock," ( )" indicate specific characteristic, quotes indicate common name	Note	Bulk Density Range (g/cm³)	Bulk Density Typical average (g/cm³)	Scale (0-5)	Streak color	Mohs hardness	UV fluorescent color SW (200-280 nm)	UV fluorescent color LW (320-400 nm)	Magnetic Properties
Acanthite	Silver sulfide	7.20	7.22	0	black	2.0-2.5			
Acmite/ Aegirine	Pyroxene (Clinopyroxene)	3.50	3.54	2	yellowish gray	6.0			
Actinolite	Amphibole group	2.95	3.04	1	white	5.0-6.0			
Adularia	Feldspar (K-spar group)	2.55	2.63	2	colorless-white	6.0			
Aeschynite-Ce	Hydrous oxide (Ce,Ca,Fe,Th,Ti,Nb) - rare earth	4.17	5.19	0	very dark brown	5.0-6.0			
Aeschynite-Y	Hydrous oxide (Y,Ca,Fe,Th,Ti,Nb) - rare earth	4.09	5.13	0	orangish yellow,	5.0-6.0			
Alabandite / alabandine	Magnesium sulfide	4.00	4.06	0	green	3.5-4.0			
Albite	Feldspar (Plagioclase group)	2.55	2.62	4	white	6.0-6.5	dull deep red		
Allanite / orthite subgroup	Sorosilicate group w/in epidote group	3.30	3.50	0	gray	5.5-6.0			
Allemontite / Stibarsen	Arsenic mineral	5.80	6.15	0	very dark gray	3.0-4.0			
Allophane	Clay: Aluminum silicate mineraloid	1.11	1.90	2	white	3.0			
Altaite	Galena group	8.10	8.14	1	black	2.5-3.0			
Alunite	Sulfate (Al,K)	2.59	2.75	1	white	3.0-4.0			
Amphibolite Group	overall	2.84	3.29	3	very light gray to medium gray	5.0-6.0			weakly-moderately
Amphibolite	Rock > 50% amphiboles	2.63	2.57	1	white	5.5-6.0			
Amblygonite	fluorophosphate	2.98	3.05	1	white	5.0-5.5			
Analcime / analcite	Silicate (Na,Al)	2.24	2.29	2	white	5.5-6.0			
Anatase	Oxide (Ti)	3.79	3.90	1	pale yellowish white	5.5-6.0			
Andalusite	Neosilicate (Al)	3.13	3.15	2	white	6.5-7.5			
Andesine	Feldspar (Plagioclase group)	2.66	2.67	4	white	6.0-6.5			
Andesite	Rock	1.99	2.78	4	white				
Andradite	Garnet group	3.70	3.86	2	white	6.5-7.0			
Anglesite	Sulfate (Pb)	6.10	6.30	1	white	2.5-3.0			
Ankerite	Carbonate group	2.93	3.05	2	white	3.5			
Annabergite	Hydrous arsenate (Ni)	3.05	3.07	1	pale green-white	1.5-2.5			
Anorthite	Feldspar (Plagioclase group)	2.72	2.74	4	white	6.0-6.5			
Anorthoclase	Feldspar (K-spar group)	2.57	2.58	4	white	6.0			
Anorthosite	Rock	2.73	2.74	2					
Anthophyllite	Amphibole group	2.85	3.20	3	very light gray to medium gray	5.5-6.0			
Anhydrite	Sulfate (Ca) - Evaporite	2.96	2.97	3	white	3.5			
Antlerite	Sulfate (Cu)	3.90	3.00	1	pale green	3.0-3.5			
Apatite	Phosphate (Ca)	3.16	3.19	3	white	5.0	deep yellow-purple (needs activator elements)	deep yellow-purple (needs activator elements)	
Apophyllite	Apophyllite group	2.30	2.34	1	white	3.5-4.0	pale yellow or pale blue (needs activator elements)	pale yellow or pale blue (needs activator elements)	
Aragonite	Carbonate group	2.91	2.93	2	white	3.5-4.0			

Mineral	Type					Color	Hardness			Notes
Arfvedsonite	Amphibole group (Na)	3.33	3.49	3.44	2	Deep blue gray-deep greenish gray	5.0-6.0			
Argillite	Rock	2.00	3.16	2.69	5					
Arsenopyrite	Sulfide (As,Fe)	5.91	6.20	6.07	1	black	5.5-6.0			
Atacamite	Halide (Cu)	3.75	3.78	3.76	1	bright green	3.0-3.5			
Augite	Pyroxene (Clinopyroxene)	3.19	3.56	3.38	5	very pale green	5.5-6.0			
Aurichalcite	Carbonate (Cu,Zn)	3.64	3.99	3.96	1	light blue	3.0			
Autunite	Hydrous phosphate (Ca,U)	3.05	3.22	3.15	1	pale yellow	2.0-2.5	yellowish green	yellowish green	
Awaruite	Alloy (Ni,Fe)	7.80	8.65	8.01	1	light gray	5.5-6.0			
Axinite	Silicate (Ca,Al,B)	3.18	3.37	3.28	1	white	6.0-7.5			
Azurite	Carbonate (Cu)	3.71	3.83	3.77	1	light blue	3.5-4.0			
Banded iron formation (BIF)	Rock	3.00	4.01	3.47	1					
Barite / barytе	Sulfate (Ba)	4.30	4.99	4.48	4	white	3.0-3.5			
Basalt (non vesicular)	Rock	2.78	3.17	2.97	5					
Basalt (vesicular)	Rock	1.88	3.06	2.77	5					
Beidellite	Phyllosilicate (clay-smectite)	2.25	2.90	2.15	5	white	1.0-2.0			
Beryl	Silicate (Be,Al)	2.63	2.92	2.90	3	white	7.5-8.0			
Biotite	Phyllosilicates (Mica group)	2.75	3.40	3.09	5	white	2.5-3.0			weakly-moderately
Bismite	Trioxide (Bi)	8.62	9.22	9.20	1	gray-black	4.5			
Bismuth	Arsenic group	9.70	9.83	9.81	2	light silver	2.0-2.5			
Boehmite / Böhmite	Bauxite component	3.02	3.05	3.03	3	white	3.5			
Borax	Evaporite	1.70	1.73	1.71	2	white	2.0-2.5			
Bornite	Sulfide (Cu,Fe) iridescent	4.90	5.31	5.09	2	nearly black	3.0-3.25			
Boulangerite	Sulfide (Pb,Sb)	5.70	6.30	6.00	1	reddish brown to nearly black	2.5-3			
Brannerite	Oxide (U,Ti)	4.23	5.43	4.82		dark yellow-greenish brown (olive)	4.5-5.5			weakly
Brochantite	Sulfate (Cu)	3.90	4.00	3.97	3	pale green	3.5-4.0			
Bronzite	Pyroxene (Orthopyroxene-Enstatite series)	3.25	3.88	3.40	3	pale brown	5.5			

Mineral or rock " / " indicates alternate name for the same mineral or rock, " ( ) " indicate specific characteristic, quotes indicate common name	Note	Bulk Density Range (g/cm³)	Bulk Density Typical average (g/cm³)	Scale (0-5)	Streak color	Mohs hardness	UV fluorescent color SW (200-280 nm)	UV fluorescent color LW (320-400 nm)	Magnetic Properties
Brookite	Oxide (Ti)	4.08 - 4.18	4.11	1	very light gray-pale yellowish gray	5.5-6.0			
Brucite	Hydroxide (Mg)	2.39 - 2.40	2.39	1	white	2.5-3.0			
Bytownite	Feldspar (Plagioclase group)	2.71 - 2.74	2.73	2	white	6.0-6.5			
Calaverite	Mtallic (Te,Au)	9.04 - 9.30	9.10	0	green to yellowish green	2.5-3.0			
Calcite	Carbonate (Ca)	2.69 - 2.73	2.71	5	colorless-white	3.0	variable colors (needs activator elements)	variable colors (needs activator elements)	
Calcio-olivine	Olivine group	2.95 - 3.97	2.99	2	colorless-white	4.5			
Calomel	Chloride (Hg)	6.45 - 7.80	7.15	2	very pale yellow	1.5			
Cancrinite	Feldspathoid group	2.42 - 2.51	2.45	2	white	5.0-6.0			
Carnallite	Halide (K,Mg)	1.50 - 1.70	1.60	3	white	2.5			
Carbonatite	Rock	2.44 - 3.19	2.87	1					
Carnotite	Vanadate (K,U,V)	3.70 - 4.70	4.65	0	bright yellow	2.0			
Celestite / celestine	Sulfate (Sr)	3.95 - 3.97	3.96	1	white	3.0-3.5	pale blue	pale blue	
Celsian	Feldspar group	3.10 - 3.39	3.25	0	white	6.0-6.5			
Cerussite	Carbonate group (Pb)	6.53 - 6.58	6.50	1	colorless-white	3.0-3.75			
Cervantite	Oxide (Sb)	6.40 - 6.60	6.50	0	pale yellow-white	4.0-4.5			
Chabazite	Zeolite group	2.05 - 2.20	2.09	0	white	4.5-5.0			
Chalcocite	Sulfide (Cu)	5.50 - 5.80	5.65	3	metallic gray-metallic black	2.5-3.0			
Chalcopyrite	Sulfide (Cu,Fe)	4.12 - 4.31	4.19	4	greenish black	3.5-4.0			
Chert	Rock	2.24 - 2.74	2.56	4					
Chlorite	Phyllosilicate (Chlorite group)	2.42 - 3.30	2.60	5	pale green-gray	2.0-2.5			paramagnetic
Chloritoid	Neosilicate (Mica group)	3.46 - 3.80	3.51	3	very light greenish gray	6.5			
Chondrodite	Neosilicate	3.10 - 3.16	3.15	1	gray	6.0-6.5	yellowish orange	orange	weakly-moderately
Chromite	Oxide (Spinel group)	4.50 - 5.09	4.80	2	brown	5.5			
Chrysoberyl	Oxide (Ba,Al)	3.50 - 3.84	3.67	1	white	8.5			
Chrysocolla	Hydrous phyllosilicate (Ca,Al)	1.90 - 2.40	2.35	2	pale bluish green	2.5-3.5 (when silica poor)			
Clay	Rock - unconsolidated	1.92 - 2.45	2.07	5					
Clay "stone"	Rock - somewhat consolidated	1.52 - 2.49	2.07	4					
Cinnabar	Sulfide	8.10 - 8.25	8.18	2	bright red	2.0-2.5			
Clinoclase / Abichite	Arsenate (Cu)	4.21 - 4.31	4.29	1	bluish green	2.5-3.0			
Clinoenstatite	Pyroxene (Clinopyroxene)	3.15 - 3.40	3.21	1	greenish gray	5.0-6.0			
Clinoferrosilite / Clinohypersthene	Pyroxene (Clinopyroxene)	3.83 - 4.10	3.74	2	white	5.0-6.25			
Clinohumite	Neosilicate	3.17 - 3.35	3.26	3	white	6.0			
Coal	Rock	1.15 - 2.09	1.69	3					
Cobaltite	Sulfide (Co,As)	6.30 - 6.50	6.33	2	dark gray-black	5.5			
Colemanite	Evaporite (B)	2.35 - 2.50	2.42	2	white	4.5	pale green		
Conglomerate	Rock - overall	2.11 - 3.00	2.57						
Conglomerate	Rock-quartz / chert dominated	2.33 - 2.79	2.59						
Copper	Native metal	8.93 - 8.95	8.94	1	metallic orange-orange red	2.5-3.0			

Cordierite / dichroite / Iolite	Cyclosilicate (Mg,Al)	2.55	2.77	2.66	1	white	7.0-7.5
Corundum	Oxide (Hematite group)	3.95	4.10	4.05	1	colorless	9.0
Covellite	Sulfide (Cu)	46.00	4.80	4.68	2	metallic dark gray	1.5-2.0
Cummingtonite	Amphibole group	3.12	3.60	3.35	2	Very pale gray	5.0-6.0
Cuprite	Oxide (Cu)	6.10	6.17	6.14	2	metallic brownish red	3.75
Dacite	Rock	2.23	2.71	2.57	2		
Diamictite (>50% quartz)	Rock	2.81	3.33	2.86	1		
Diamictite (>50% feldspars)	Rock	2.78	2.95	2.80	1		
Diamictite (>50% Lithics)	Rock	2.89	4.79	3.01	1		
Diamond	Native mineral	3.51	3.53	3.52	1	colorless	10.0
Diabase / Dolerite	Rock	2.97	3.10	2.97	2		
Diorite	Rock	2.55	3.17	2.75	3		
Dickite	Phyllosilicate (clay)	2.50	2.70	2.60	4	white	1.5-2.0
Digenite	Sulfide (Cu)	5.50	5.70	5.63	2	black	2.5-3.0
Diopside	Pyroxene (Clinopyroxene)	3.20	3.40	3.28	3	white	5.5-6.0
Dolomite	Carbonate (Ca,Mg)	2.84	2.86	2.84	3	white	3.5-4.0
Donpeacorite	Pyroxene (Orthopyroxene)	2.24*	3.40	3.36	3	colorless (breaks into splinters)	5.0-6.0
Dolostone (pure)	Rock	2.81	2.90	2.84	4		
Dolostone (argillaceous)	Rock	2.78	2.91	2.83	4		
Dumortierite	Neosilicate (Al)	3.30	3.40	3.34	2	white	7.5-8.0
Edenite	Amphibole (Hornblende series)	3.02	3.37	3.05	3	white	5.0-6.0
Enstatite	Pyroxene (Orthopyroxene)	3.20	3.30	3.22	1	gray	5.0-6.0
Epidote	Epidote group	3.38	3.49	3.43	5	Very pale gray	6.0-6.75	weakly-moderately
Epsomite	Evaporite (Sulfate (Mg))	1.67	1.68	1.67	4	white	2.0

255

Mineral or *rock* "/" indicates alternate name for the same mineral or rock," ( )" indicate specific characteristic, quotes indicate common name	Note	Bulk Density			Scale (0-5)	Streak color	Mohs hardness	UV fluorescent color		Magnetic Properties
		Range (g/cm³)	Typical average (g/cm³)					SW (200-280 nm)	LW (320-400 nm)	
Essenelte	Pyroxene (Clinopyroxene)	3.50	3.60	3.54	3	white	6.0			
Eudlese	Neosilicate	2.99	3.10	3.04	2	white	7.5			
Eulite (see *Ferrosilite*)										
Euxenite	Oxide (Y,Ca,Ce,U,Th)	4.70	5.00	4.84	0	yellowish gray, reddish brown	5.5-6.5			
Fayalite	Olivine group	4.35	4.43	4.39	3	white	6.5-7.0			
Fergusonite	Rare earth oxide (Y,Nb)	4.50	5.75	5.05	0	brown	5.5-6.0			
Ferrosilite	Pyroxene (Orthopyroxene-Enstatite group)	3.65	3.95	3.88	3	pale grayish brown	5.0-6.0			
Fluorite	Halide	3.13	3.18	3.56	4	white	4.0	varies; usually pale blue		
Forsterite	Olivine group (end member)	3.21	3.33	3.27	4	white	7.0			
Franklinite	Oxide (Zn,Fe)	5.07	5.22	5.14	2	reddish brown to nearly black	5.5-6.0			moderately-strongly
Gabbro	Rock	2.79	3.34	2.98	4					
Gadolinite / ytterbite	Rare earth oxide (Ce,La,Nd,Y)	4.00	4.50	4.25	0	greenish gray	7.0			weakly
Gahnite	Oxide (Spinel group)	4.00	4.60	4.38	1	gray	7.5-8.0			
Galaxite	Oxide (Spinel group)	4.20	4.30	4.23	1	reddish brown	7.5			weakly-moderately
Galena	Sulfide (Pb)	7.20	7.60	7.40	4	dull metallic gray	2.5-2.75			
Garnet Group	overall	3.50	4.30	variable	3	usually white	6.25-7.75			non-weakly
Gersdorffite	Sulfide (Ni,As)	5.90	6.33	6.11	2	dark gray-black	5.0-5.5			
Gibbsite / hydrargillite	Hydroxide (Al)	2.30	2.40	2.34	3	white	2.25-3.0			
Glauberite	Sulfate (Na,Ca)	2.75	2.85	2.77	3	white	2.5			
Glaucochroite	Silicate (Ca,Mn)	3.40	3.45	3.41	2	white	6.0			
Glauconite	Phyllosilicate (Mica group)	2.42	2.95	2.69	2	light green	2.0			
Glaucophane	Amphibole group	3.00	3.15	3.07	3	grayish blue	6.0-6.5			
Gmelinite	Zeolite group	2.04	2.17	2.10	0	white	4.5			
Gneiss	Rock	1.83	3.15	2.81	5					
Goethite	Oxyhydroxide (Fe)	3.30	4.30	4.85	2	brown to yellowish brown	5.0-5.5			weakly
Gold	Native metal	19.30	19.34	19.32	0	metallic yellow	2.5			
Granite	Rock	2.53	2.82	2.64	5					
Granodiorite	Rock	2.56	2.79	2.64	3					
Granulite	Rock	2.63	3.00	2.77	2					
Graphite		1.90	2.30	2.16	3	black	1.0-2.0			
Green sandstone	Rock	2.21	2.52	2.37	3					
Grossularite / grossular (gooseberry garnet)	Garnet group	3.42	3.72	3.61	3	brown	6.5-7.0	pale orange	pale yellow orange	
Gypsum	Hydrous sulfate (Ca)	2.30	2.33	2.31	4	white	1.75-2.0	red or orange (when Mg present)	will fluoresce when dissolved in tonic water	
Halite (table salt)	Halide	2.15	2.20	2.17	5	white	2.0-2.5			

Halloysite	Aluminosilicate (clay)	2.00	2.65	2.60	2	lighter than mineral color-white	2.0-2.5	
Harzburgite	Rock (type of peridotite)	3.00	3.50	3.33	2			
Hausmannite	Oxide (Mn)	4.70	4.84	4.76	1	dark reddish brown	5.5	
Hectorite	Phyllosilicate (clay-smectite)	2.00	3.00	2.50	0	white	1.0-2.0	
Hedenbergite	Pyroxene (Clinopyroxene)	3.55	3.57	3.56	2	very pale gray-gray	5.5-6.0	
Hematite	Oxide (Fe)	5.26	5.31	5.30	3	bright red to dark red	5.5-6.5	
Hercynite	Oxide (Spinel group)	3.95	3.95	3.95	1	dark green	7.5	
Hessite	Telluride (Ag)	7.20	8.20	7.90	0	black	2.0-3.0	
Hornblende minerals	Inosilicate amphiboles	2.90	3.47	3.12	4	colorless-light gray	5.0-6.0	
Huebnerite	Oxide (Mn,W) Wolframite group-end member	7.12	7.18	7.15	0	dirty yellow-greenish gray	4.0-4.5	
Humite	Humite group	3.15	3.32	3.19	0	white	6.0-6.5	
Hyalophane	K-spar (similar to orthoclase and adularia)	2.75	2.87	2.81	1	colorless-white	6.0-6.5	
Hydrozincite / marionite (zinc bloom)	Carbonate (Zn)	3.50	3.85	3.68	1	white	2.0-2.5	
Hypersthene	Pyroxene (orthopyroxene)	3.20	3.90	3.40	3	pale gray-pale greenish gray	5.5-6.0	
Ice	(common terrestrial "water" ice)	0.90	0.98	0.92	5	colorless	1.25-1.5 (at freezing point)	
Idocrase / vesuvianite	Sorosilicate	3.32	3.43	3.39	2	white	6.5	
Ignimbrite / welded tuff	Rock	1.24	2.62	2.13	2			
Illite , hydromica	Mica phyllosilicates (clay) closely related group of minerals	2.60	2.90	variable	4	white	1.0-2.0	
Ilmenite	Oxide (Ti,Fe)	4.69	4.79	4.72	2	black	5.0-6.0	weakly-moderately
Ilvaite / yenite	Sorosilicate	3.80	4.09	4.04	2	black	5.5-6.0	
Immature quartzite	Rock	2.45	2.89	2.67	3			
Iodyrite / iodargyrite	Halide (Ag)	5.60	5.73	5.69	1	metallic yellow	1.5-2.0	
Iridium	Native metal (used to refer to any Pt group metals or combo of the Pt group of metals)	22.50	22.80	22.56	1	white-light gray	6.5-7.0	moderately (under STP)
Iron	0	7.85	7.89	7.87	2	metallic medium gray	4.0	
Jacobsite	Oxide (Mn) (Spinel group)	4.73	4.76	4.75	2	dark reddish brown nearly black	5.5-6.0	

Mineral or rock "/" indicates alternate name for the same mineral or rock." ()" indicate specific characteristic, quotes indicate common name	Note	Bulk Density Range (g/cm³)	Bulk Density Typical average (g/cm³)	Scale (0-5)	Streak color	Mohs hardness	UV fluorescent color SW (200-280 nm)	UV fluorescent color LW (320-400 nm)	Magnetic Properties
Jadeite	Pyroxene (Clinopyroxene)	3.24	3.31	3	white	6.5-7.0		very pale green	
Jamesonite	Sulfide (Pb,Fe,Sb), Jamesonite group	5.56	5.67	3	dark gray-black	2.5			
Jervisite	Pyroxene (Clinopyroxene)	2.24*	3.33*	2	white	6.0			
Johannsenite	Pyroxene (Clinopyroxene)	3.20	3.56	3	white	6.0			
kanoite	Pyroxene (Clinopyroxene)	3.61	3.66	2	light brown	6.0			
Kaolinite	Phyllosilicate (Kaolinite group) clay	2.16	2.61	5	white	2.0-2.5			
Kirschsteinite / kirschsteinite / Monticellite	Olivine group	3.05	variable	2	white	5.0-5.5			
kosmochlor	Pyroxene (Clinopyroxene)	3.51	3.56	1	light green	6.0			
Kyanite / cyanite	Silicate (Al)	3.59	3.66	2	white	4.5-parallel to long axis, 6.5-7.0 perpendicular to axis			
Labradorite	Feldspar (Plagioclase group)	2.68	2.70	1	white	6.0-6.5			
Lamprophyre	Rock	2.65	2.93	1					
Langbeinite / K-Mag / Sul-Po-Mag	Sulfate (K,Mg)	2.87	2.83	1	white	3.5-4.0			
Lathinite	Olivine group	3.90	3.97	2	pale brown-dark brown	5.5-6.5			moderately
Larseenite	Olivine group	5.90	5.90	2	colorless-white	3.0			
Latite	Rock	1.98	2.80	3					
Laumonite	Zeolite group	2.25	2.29	0	white	4.0			
Lawsonite	hydrated variant of "anorthite"	3.05	3.09	2	white	7.5			
Lazurite	Tectosilicate (sodalite group)	2.38	2.40	2	bright blue	5.0-5.5			
Lechatellerite	Mineraloid of SiO₂	2.50	2.60	1	white	6.5			
Lepidolite	Phyllosilicate (Mica group)	2.80	2.84	1	white	2.5-3.0			
Leucite	Feldspathoid group	2.45	2.47	1	white	5.5			
Lherzolite	Rock (type of peridotite)	3.22	3.25	2					
Liebenbergite	Olivine group (Ni)	3.22*	4.22	1	white	6.0-6.5			
Limonite	Mineraloid: oxides (fine grained, Fe)	2.70	variable	4	yellowish brown	4.0-5.5			weakly-moderately
Limestone (pure)	Rock	1.59	2.48	5					
Limestone (argillaceous)	Rock	1.64	2.59	5					
Loadstone (see magnetite)	Rock (magnetite aligned to magnetic north)								
Linnaeite	Sulfide (Co)	4.80	4.85	1	dark gray	4.5-5.5			strongly
Loellingite	Arsenate (Fe)	7.10	7.40	1	dark gray	5.0-5.5			
Maghemite	Oxide (Fe)	4.86	4.90	1	brown	5.0			strongly
Magnesite	Carbonate group (Mg)	3.00	3.10	3	white	3.5-4.5			
Magnetite	Oxide (spinel group)	5.15	5.17	4	black	5.5-6.5	pale blue	pale blue	strongly
Malachite	Carbonate (Cu)	3.60	3.87	2	light green	3.5-4.0			
Manganite	Oxide (Mn)	4.29	4.31	2	reddish brown to nearly black	4.0			
Manganosite	Oxide (Mn)	5.18	5.37	1	brown	5.0-6.0			
Marble	Rock	2.51	2.72	3					
Marcasite	Sulfide (Fe)	4.88	4.89	3	dark gray-black	6.0-6.5			

Margarite	Phyllosilicate (Mica group)	2.99	3.08	3.03	3	white	4.0
Marialite	Tectosilicate (scapolite group)	2.50	2.62	2.56	1	white	5.5-6.0
Meionite	Tectosilicate (scapolite group)	2.69	2.78	2.74	1	white	5.0-6.0
Microcline	Feldspar group (K-spar)	2.51	2.60	2.56	3	white	6.0-6.5
Microlite	Oxide (Na,Ta)	4.20	6.40	5.30	2	pale yellow to brown	5.5
Migmatite	Rock	2.72	3.13	2.74	1		
Minium	Oxide (Pb)	8.20	9.10	8.50	2	yellow orange	
Molybdenite	Sulfide (Mo)	4.62	5.00	4.73	1	bluish gray	1.0-1.5
Monazite group (overall)	Phosphate (Ce,La,Th) rare earth	4.60	5.40	5.18	0	white	5.0-5.5
Monticellite (see *Kerschsteinite*)	Olivine group						
Montmorillonite	Phyllosilicate (clay-smectite)	2.00	2.70	variable	3	white	1.0-2.0
Monzonite	Rock	2.79	2.90	2.85	4		
Mozartite	Hydrous silicate (Ca,Mn)	3.63	3.65	3.64	1	red	6.0
Mudstone (silt / clay mix)	Rock	1.32	2.87	2.20	5		
Mullite / porcelainite	Neosilicate (Al) "this data is for the mineral, not impure chert"	2.60	*3.10*	3.05	2	white to pale colors	5.5-6.75
Muscovite	Phyllosilicate (mica group)	2.76	3.01	2.82	4	white	2.0-2.5 (on platy surface), 4.0 (on platy edges)
Nacrite	Phyllosilicate (kaolinite serpentine group) clay	2.16*	2.69*	2.60	3	white	2.0-2.5
Nagyagite	Sulfide (Pb,Au,Te,Sb)	7.39	7.49	7.49	0	metallic gray-metallic black	1.5
Numansilite	Pyroxene (Clinopyroxene)	3.50	3.61	3.51	1	brownish red	6.0-7.0
Natalyite	Pyroxene (Clinopyroxene) (Cr,V)	?	?	3.55	1	green	7.0
Natrolite	Zeolite group	2.20	2.26	2.25	1	white	5.0-5.5

Mineral or rock "/" indicates alternate name for the same mineral or rock," ()" indicate specific characteristic, quotes indicate common name	Note	Bulk Density Range (g/cm³)	Bulk Density Typical average (g/cm³)	Scale (0-5)	Streak color	Mohs hardness	UV fluorescent color SW (200-280 nm)	UV fluorescent color LW (320-400 nm)	Magnetic Properties
Nchwaningite	Pyroxene (Orthopyroxene)	2.24*	3.20	0	white	5.5			
Nepheline	Feldspathoid group	2.55 - 2.65	2.59	1	white	6.0			
Nepheline clinopyroxene / Ijolite	Rock	2.99 - 3.02	2.99	1					
Nepheline monzogabbro / Essexite	Rock	2.94 - 3.04	2.98	1					
Nepheline syenite	Rock	2.64 - 2.70	2.65	1					
Nickeline / niccolite	Arsenate (Ni)	7.78 - 7.80	7.79	1	very dark brown	5.0-5.5			
Nitre / niter	Oxide (K,N)	2.10 - 2.10	2.10	1	white	2.0			
Nontronite	Phyllosilicate (clay-smectite)	2.00* - 2.60*	2.30	2	colorless	1.5-2.0			
Norite	Rock	2.98 - 3.06	3.02	1					
Nosean / noselite	Feldspathoid group	2.21 - 2.40	2.30	3	very pale blue	5.5			
Oligoclase	Feldspar (Plagioclase group)	2.63 - 2.66	2.65	4	white	6.0-6.5			
Olivine	Olivine group	3.22 - 4.13	3.37	3	colorless-white	6.5-7.0			
Omphacite	Pyroxene (Clinopyroxene)	3.29 - 3.39	3.14	3	very pale green	5.0-6.0			
Opal	Mineraloid of hydrous SiO₂	1.25 - 2.23	2.09	2	white	5.5-6.0	white with hints of yellow, green, or blue		
Orpiment	Sulfide (As)	3.49 - 3.55	3.52	2	pale bright yellow	1.5-2.0			
Orthoclase	Feldspar (K-spar group)	2.52 - 2.63	2.56	4	white	6.0			
Ottrelite (chloritoid variety)	Neosilicate (Mica group)	3.50 - 3.55	3.52	1	greenish gray	6.0-7.0			
Pargasite	Phyllosilicate (Mica group)	2.76 - 2.78	2.78	2	white	2.5-3.0			
Pargasite	Amphibole (Hornblende series)	3.07 - 3.18	3.12	2	pale grayish green	5.0-6.0			
Pearceite	Sulfide (Cu,Ag,As) "Ruby silver"	6.10 - 6.15	6.13	1	black	3.0			
Pectolite	Inosilicate hydroxide (Na,Ca); primary mineral in nepheline syenites	2.84 - 2.90	2.86	3	white	4.5-5.0			
Pentlandite	Sulfide (Fe,Ni)	4.60 - 5.00	4.65	2	light bronze	3.5-4.0			
Peridotite	Rock	2.97 - 3.35	3.34	1					
Perovskite	Oxide (Ca,Ti)	3.98 - 4.26	4.01	2	very light gray	5.0-5.5			
Petalite / castorite	Feldspathoid group	2.39 - 2.46	2.42	2	colorless	6.0-6.5			
Petedunnite	Pyroxene (Clinopyroxene)	3.30 - 3.68	3.35	1	very pale green	5.0-6.0			
Petzite	Telluride (Ag,Au)	8.70 - 9.14	8.92	1	very dark gray	2.5-3.0			
Phenakite (be)	Neosilicate (be)	2.97 - 3.00	2.98	1	white	7.5-8.0			
Phenakitte	Phyllite	2.68 - 2.80	2.74	4					
Phillipsite	Zeolite group	2.19 - 2.20	2.20	0	white	4.0-4.5			
Phlogopite	Phyllosilicate (Mica group)	2.78 - 2.85	2.83	3	white	2.0-2.5			
Phonolite	Rock	2.37 - 2.63	2.54	2					
Phosgenite	Carbonate chloride (Pb)	6.00 - 6.30	6.15	1	white	2.0-3.0			
Pigeonite	Pyroxene (Clinopyroxene)	3.30 - 3.46	3.38	3	very light gray	6.0			
Plagionite	Sulfide (Pb,Sb), Jamesonite group	5.40 - 5.60	5.56	2	dark metallic gray	2.5			
Platinum	Native metal	21.42 - 21.47	21.45	0	metallic light gray	4.0-4.5			

Name	Type					Color	Hardness			Notes
Polybasite	Sulfate (Ag,Cu,Sb,As)	4.60	5.00	4.80	1	black with a red tint	1.5-2.0			
Polybasite-Tac	Sulfate (Ag,Cu,Sb,As)	6.33	6.35	6.34	1	black	3.0			
Polycrase (polycrase-Y)	Oxide-rare earth (Y,Co,Ce,U,Th,Ti,Nb,Ta)	4.70	5.90	5.00	0	near black with red, orange, or yellow tints	5.0-6.0			
Powellite	Molybdate (Ca)	4.25	4.50	4.34	1	light yellow	3.5-4.0	bright yellow	pale yellow	
Prehnite	Inosilicate (Ca,Al)	2.80	2.95	2.87	3	white	6.0-6.5		pale pinkish orange	
Protoquartzite	*Rock*	2.50	2.72	2.57	3					
Proustite	Sulfate (Ag,As)	5.50	5.60	5.55	1	bright red	2.0-2.5			
Psilomelane group	Oxide (Mn)	4.40	4.72	4.56	2	very dark brown	5.0-6.0			non-moderately
Pumice	*Rock*	1.15	2.23	1.79	4					
Pyrargyrite "ruby silver"	Sulfate (Ag,Sb)	*5.05*	5.86	5.19	1	deep cherry red	2.5			
Pyrite	Sulfide (Fe)	4.85	5.07	5.01	4	greenish to brownish very dark gray	6.0-6.5			paramagnetic
Pyrochlore	Oxide (Na,Ca,Nb)	*4.20*	*6.40*	*5.30*	2	white	5.0-5.5			
Pyrolusite	Oxide (Mn)	4.80	5.23	5.06	3	black with a blue tint	6.0-6.5 (when crystalline)			
Pyromorphite	Phosphate (Pb)	6.70	7.00	6.85	3	white	3.5-4.0	light yellow-orange	light yellow-orange	
Pyrope	Garnet group	3.65	3.84	3.74	2	white	7.0-7.5			
Pyrophyllite	Silicate hydroxide (Al)	2.81	2.89	2.84	3	white	1.5-2.0			
Pyroxenite (overall)	*Rock*	2.24	3.33	3.28	2					
Pyrrhotite	Sulfide (Fe)	4.58	4.65	4.61	3	gray-very dark gray	3.5-4.5			weakly (strongly: if monoclinic)
Quartz	euhedral	2.63	2.66	2.65	5	colorless	7.0			
Quartz	anhedral (rock)	2.59	2.65	2.64	5	colorless-white	7.0			
Quartzite	*Rock*	2.51	2.77	2.63	4					
Quartz diorite	*Rock*	2.81	2.86	2.82	4					
Quartz monzonite	*Rock*	2.70	2.73	2.71	4					

Mineral or rock "/" indicates alternate name for the same mineral or rock," ( )" indicate specific characteristic, quotes indicate common name	Note	Bulk Density Range (g/cm³)	Bulk Density Typical average (g/cm³)	Scale (0-5)	Streak color	Mohs hardness	UV fluorescent color SW (200-280 nm)	UV fluorescent color LW (320-400 nm)	Magnetic Properties
Realgar "ruby sulfur, ruby arsenic"	Sulfide (As)	3.56 3.59	3.57	1	reddish orange-red	1.5-2.0			
Rhodochrosite	Carbonate (Mn)	3.69 3.70	3.69	3	white	3.5-4.0			
Rhodonite	Inosilicate	3.40 3.76	3.50	2	white	5.5-6.5			
Rhyolite (massive)	Rock	2.63 2.91	2.70	3		6.0			
Riebeckite	Amphibole group	3.40 3.40	3.40	3	pale bluish gray-bluish gray				
Roscoelite	Phyllosilicate (Mica group)	2.93 2.94	2.84	3	gray	1.0-2.5			
Rutile	Oxide (Ti)	4.23 4.25	4.24	2	bright red-dark red	6.0-6.5			
Sanidine	Feldspar (K-spar) high temperature	2.52 2.62	2.56	4	white	6.0			
Sand	Rock (unconsolidated)	1.69 3.22	2.05	5					
Sandstone (arenite)	Rock	2.03 2.99	2.53	5					
Sandstone (wacke)	Rock	1.88 2.87	2.37	5					
Saponite	Phyllosilicate (clay-smectite)	2.24 2.31	2.30	3	white	1.5			
Schist	Rock	2.39 2.94	2.70	4					
Scolecite	Zeolite group	2.16 2.40	2.33	0	white	5.0-5.5			
Scorodite	Hydrous arsenate (Fe)	3.22 3.28	3.27	1	very pale green	3.5-4.0			
Semseyite	Sulfate (Pb,Sb)	5.80 6.19	6.08	1	black	2.5			
Serpentine subgroup	overall	2.53 2.65	2.60	3	pale green	2.5-4.0			
Serpentinite	Rock	2.24 3.27	2.63	1					
Shale	Rock	1.61 2.73	2.34	5					
Silt	Rock (unconsolidated)	1.72 2.16	1.98	4					
Siltstone	Rock	2.28 2.99	2.55	4					
Siderite / chalybite	Carbonate (Fe)	3.93 4.00	3.96	3	white	3.75-4.25			
Sidevite	Rock	2.65 2.73	2.67	3					
Sillimanite	Neosilicate (Al)	3.13 3.65	3.24	1	white	7.0			
Silver	Native metal	10.47 10.52	10.49	0	metallic, light gray	1.5			
Slate	Rock	2.71 2.85	2.77	3					
Smectite	Clay	2.00 2.60	2.35	3	white	1.0-2.0			
Smithsonite / calamine / "zinc spar"	Carbonate (Zn)	4.40 4.50	4.45	2	white	4.5	Red (needs activator elements)	Red (needs activator elements)	
Sodalite	Feldspathoid group	2.27 2.33	2.29	2	white	5.5-6.0	Blaze orange	Blaze orange	
Sperrylite	Arsenide (Pt)	10.50 10.80	10.58	0	black	6.0-7.0			
Spessartine	Garnet group	4.05 4.40	4.17	1	white	6.5-7.5			
Sphalerite	Sulfide (Zn,Fe)	3.90 4.10	4.05	1	pale yellow-very light brown	3.5-4.0	blues-purple, bright orange	same as short wave but duller	
Sphene / titanite	Neosilicate (Ca,Ti)	3.45 3.60	3.48	1	very pale red or light pink	5.0-5.5			
Spinel group	overall	3.50 3.61	3.58	2	highly variable	5.0-8.0			non-strongly
Spodumene	Pyroxene (Clinopyroxene)	3.15 3.20	3.17	3	white	6.5-7.0			

Staurolite	Neosilicate (Fe,Al) forms crosses	3.65	3.83	3.76	2	white-light gray	7.0-7.5		paramagnetic
Sternbergite	Sulfide (Ag,Fe)	4.10	4.28	4.22	1	black	1.0-1.5		
Stibnite / antimonite / "kohl"	Sulfide (Sb)	7.10	7.20	7.12	3	Metallic gray	2.0		
Stilbite	Zeolite group	2.12	2.22	2.15	2	white-very light pink	3.5-4.0		
Stillwellite-Ce	Borosilicate (Ce,La,Ca) rare earth	4.16	4.61	4.59	0	white	6.5		
Stolzite	Tungstate (Pb)	7.90	8.20	8.05	1	white	2.5-3.0		
Strontianite	Carbonate group	3.70	3.80	3.78	2	white	3.5		
Sulphur	Native element	1.92	2.07	2.06	2	pale yellow	1.5-2.5		
Syenite	Rock	2.57	2.75	2.62	4				
Sylvanite	Telluride (Ag,Au)	6.90	8.56	6.97-8.10	1	metallic gray	1.5-2.0		
Sylvite / potassium salt	Halide (K)	1.96	1.99	1.98	3	white	1.5-2.0		
Talc	Hydrated silicate (Mg)	2.60	2.80	2.75	4	colorless-white	1.0	orangish yellow	yellow
Tantalite	Oxide (Ta,Fe,Mn) rare earth	6.20	8.00	6.65	0	deep reddish brown-black	6.0-6.5		
Tennantite	Sulfide (Cu,As)	4.60	4.70	4.66	1	reddish gray	3.0-4.5		
Tephroite	Olivine group	3.81	4.25	3.81	2	very light gray	6.0		
Tetrahedrite	Sulfide (Sb,Cu,Fe)	4.60	5.15	4.94	1	deep red-black	3.5-4.0		
Thenardite	Sulfate (Na)	2.67	2.70	2.68	2	white	2.5	bright white	yellowish green
Tonalite	Rock	2.57	2.95	2.70	2				
Thorianite	Oxide (Th)	9.50	10.00	9.70	1	gray-greenish gray	6.5-7.0		
Thorite	Silicate (U,Th)	4.00	6.70	5.35	2	light orange-light brown	4.0-4.5		
Tin	Native metal	7.28	7.37	7.31	2	very light metallic gray	1.5		paramagnetic strongly
Titanomagnetite	Oxide (Ti,Fe)	4.80	5.30	5.10	2	black	5.0-5.5		
Tourmaline group	overall	2.82	3.32	3.07	4	white	7.0-7.5	pale red to pale purple	pale red to pale purple
Trachyte	Rock	2.17	2.95	2.59	2				
Tremolite	Amphibole group	2.99	3.04	3.01	3	white	5.0-6.0	dull red-dull yellow	
Topaz	Hydrous neosilicate (Al,F)	3.43	3.59	3.54	1	white	8.0	golden yellow	milky white

Mineral or *rock* "/" indicates alternate name for the same mineral or rock," ( )" indicate specific characteristic, quotes indicate common name	Note	Bulk Density Range (g/cm³)	Bulk Density Typical average (g/cm³)	Scale (0-5)	Streak color	Mohs hardness	UV fluorescent color SW (200-280 nm)	UV fluorescent color LW (320-400 nm)	Magnetic Properties
Tridymite	Quartz group	2.25-2.33	2.28	2	white	7.0	deep red		
Triphylite	Phosphate (Li,Fe)	3.47-3.58	3.50	2	very light gray	4.0-5.0			
Trona	Hydrated carbonate (Na)	2.11-2.17	2.13	2	white	2.5			
Tuff	Rock	1.41-2.94	2.11	4					
Tungstite	Hydrated oxide (W)	5.50-5.68	5.52	1	yellow	2.5			
Tyuyamunite	Hydrous (U,V,)	3.36-4.30	3.80	0	yellow	1.5-2.0			
Ultramafic (overall)	Rock	2.71-4.13	3.29	2					
Uraninite / pitchblende	Oxide (U)	7.20-10.95	8.72	1	dark olive-dark brown	5.0-6.0			
Uranophane / uranolite	Hydrated silicate (Ca,U)	3.81-3.90	3.86	0	pale yellow	2.5	faint yellowish green	faint yellowish green	
Uvarovite	Garnet group (Cr)	3.35-3.81	3.77	3	white	6.5-7.5	red	red	
Vanadinite	Apatite group (Pb,V)	6.80-7.10	6.94	1	brownish yellow	3.0-4.0			
Vermiculite	Hydrated phyllosilicate (Mica group)	2.40-2.70	2.50	3	variable	1.5-2.0			
Vivianite	Sulfide (Fe,Ni)	4.70-4.80	4.79	1	black	4.5-5.5			
Vivianite	Hydrated phosphate (Fe)	2.60-2.70	2.68	2	deep blue-greenish blue	1.5-2.0			
Webstreite	Rock (type of pyroxenite)	3.30-3.26	3.19	2					
Wehrlite	Rock (type of peridotite)	3.37-4.10	3.22	2					
Willemite	Silicate (Zn)	3.87-4.10	4.06	3	variable	5.5	green	green	
Witherite	Carbonate group (Ba)	4.22-4.31	4.29	2	white	3.0-3.5	light blue	light blue	
Wolframite	Oxide (W,Fe,Mn)	7.10-7.50	7.30	2	reddish brown	4.0-4.5			
Wollastonite	Inosilicate (Ca)	2.84-3.09	2.86	3	white	4.5-5.0	deep yellow	orangish yellow	
Wulfenite	Molybdate (Pb)	6.50-7.00	6.75	2	white	3.0			
Xenotime	Phosphate (Y) rare earth	4.45-5.10	4.80	1	white	4.25			
Zeolite Group	(overall)	2.20-2.44	2.33	2	white-very light pink	3.0-5.5			
Zincite	Oxide (Zn,Mn)	5.43-5.68	5.56	0	yellowish orange	4.0			
Zinkenite	Jamesonite group	5.12-5.35	5.23	3	metallic gray	3.0-3.5			
Zircon	Neosilicate (Zr)	4.58-4.69	4.64	4	white	7.5	dull yellow	dull orange	
Zoisite	Epidote group	3.15-3.36	3.30	2	colorless-white	6.0-7.0			
List average mean	overall list	3.78-4.05	3.85		white	4.75-5.0			
List average mean	overall of earth's crust	2.23-2.86	2.57		colorless-white	5.75-6.0			

LEGEND

	= Data based off personally conducted lab testing
No fill	= possible radioactive hazard (only colored on minerals that can pose a radioactive hazard to people and pets)
3.0	= Data based off theoretical parameters and other worker's data
	= Mohs hardness defining mineral
3.33*	= The value has never been calculated or measured so the overall group/rock type was used
Sandstone	= Rock
Italic numbers	= Estimated number
STP	= Standard temperature and pressure (0°C or 273.15K and 1 atmosphere or 1.2754 kg/m³ of pressure)

NOTES:

Only minerals have streak colors. Rocks do not.

"Scale (0-5)" refers to the relative commonality of a rock or mineral on a global scale. 0=very rare, 1=rare, 2=rare to somewhat rare, 3=somewhat common, 4=common, 5=very common: *Just because something may be a "5" globally doesn't necessarily mean it's a "5" by your location.*

Blank cells are "not applicable"

This chart is a combination of the author's own work and the works in the references below. An online artificial intelligence (AI) generator was only used in the rare case when no other information could be located.

The "typical average (g/cm³)" is not exactly a mean average of all samples for a material. It is the mean average of the top several modal occurrences.

Ultraviolet (UV) light on the chart, is divided into short wave (SW) and long wave (LW). Middle wave (MW) is not on the chart. It's not easy to get lights in that range. SW = 200 to 280 nm, MW = 280 to 315 nm, LW = 320 to 400 nm 365 nm UV light works the best for LW UV.

The 2nd to last row in light gray, are the mean averages from this list only. The "bulk density" columns are the mean average of every number that column. Wasn't incorporated to adjust bulk density. Streak color is the most common color on the list. "Mohs hardness" is the mean average of the highest value if a range is given. A range in hardness is given because the mean =4.82. "UV fluorescent color" and "magnetic properties" have no mean average values. The values are for information only. They do not accurately reflect the earth's crust as a whole.

The last row in light grayish green, are the mean averages from the earth as a whole calculated by the author. The "bulk density" columns are the mean average of calculations done in 2018. Streak color is the most commonly occurring streak color. "Mohs hardness" is the mean average range of about 1500 minerals. A range in hardness is given because the mean =5.87. "UV fluorescent color" and "magnetic properties" have no mean average values.

REFERENCES:

Anthony, J.W., Bideaux, R.A., Bladh, K.W., Nichols, M.C., (Eds.), 2023. Handbook of Mineralogy, Mineralogical Society of America, Chantilly, VA 20151-1110, USA. handbookofmineralogy.org/

Chesterman, C.W., 2022 (36th edition) 1979 (1st edition). National Audubon Society, Field guide to rocks and minerals. Chantileer Press, Inc.

Crawford, K.M., 2013. Determination of bulk density of rock core using standard industry methods. Michigan Technological University, Master's thesis

Skladzrein, P.B., 2007. Compilation of rock densities for Victoria. GeoScience Victoria Gold Undercover Report, Department of Primary Industries

Daly, R.A., 1935. Densities of rocks calculated from their chemical analyses. Proceedings of the National Academy of Sciences (PNAS), v.23

Crystal Systems

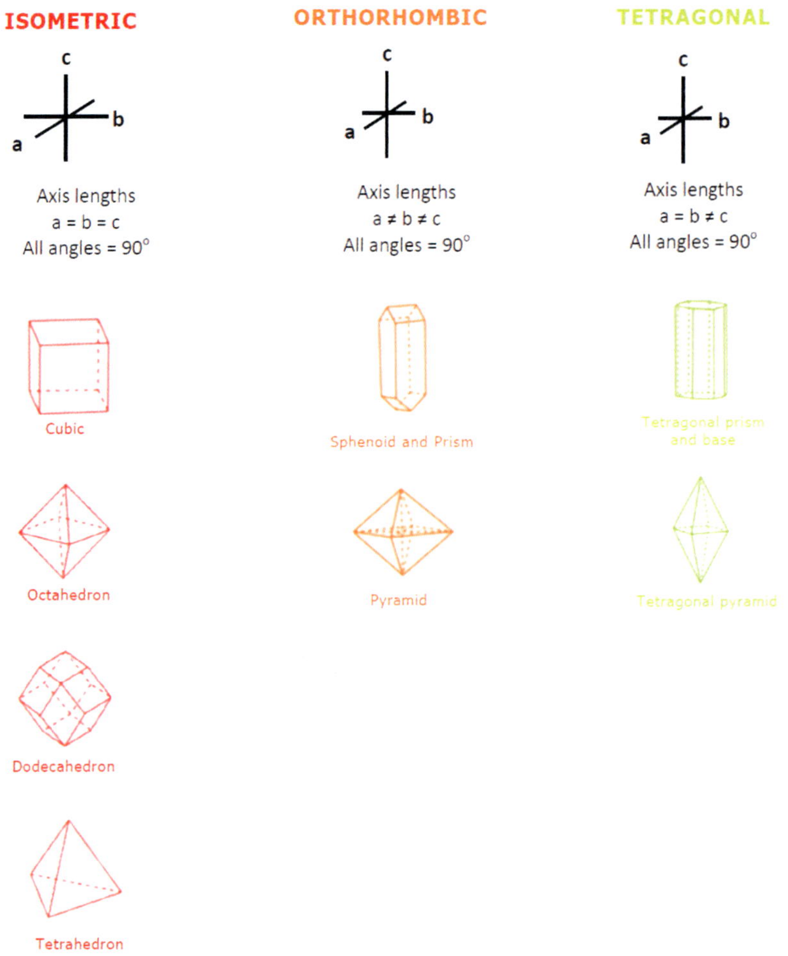

ISOMETRIC

Axis lengths
a = b = c
All angles = 90°

Cubic

Octahedron

Dodecahedron

Tetrahedron

ORTHORHOMBIC

Axis lengths
a ≠ b ≠ c
All angles = 90°

Sphenoid and Prism

Pyramid

TETRAGONAL

Axis lengths
a = b ≠ c
All angles = 90°

Tetragonal prism and base

Tetragonal pyramid

The International Mineralogical Association (IMA) is the organization that is responsible for reviewing new and old minerals.

Minerals are natural, homogeneous and inorganic crystalline solids, constituting rocks, characterized by a defined chemical composition and a particular crystallographic structure that are bound by planar surfaces.

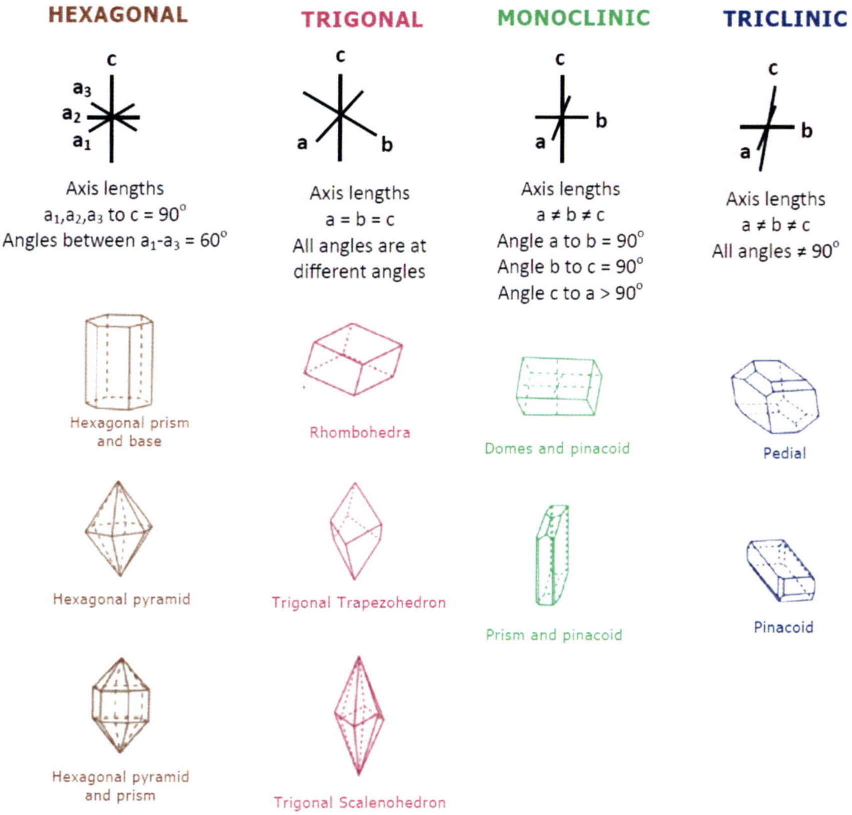

HEXAGONAL

Axis lengths
a_1, a_2, a_3 to c = 90°
Angles between a_1-a_3 = 60°

Hexagonal prism and base

Hexagonal pyramid

Hexagonal pyramid and prism

TRIGONAL

Axis lengths
a = b = c
All angles are at different angles

Rhombohedra

Trigonal Trapezohedron

Trigonal Scalenohedron

MONOCLINIC

Axis lengths
a ≠ b ≠ c
Angle a to b = 90°
Angle b to c = 90°
Angle c to a > 90°

Domes and pinacoid

Prism and pinacoid

TRICLINIC

Axis lengths
a ≠ b ≠ c
All angles ≠ 90°

Pedial

Pinacoid

All the crystal systems above and on the previous page, encompass all minerals. Every known mineral fits into one of these seven systems. Individual systems are defined by their 3D axis relationships, not by their apparent shape.

When identifying a crystal in hand specimen, it always fits within the simplest system. The systems from simplest to most complex are isometric, orthorhombic, tetragonal, hexagonal / trigonal, monoclinic, and triclinic.

NOTE: Trigonal use to be in with hexagonal, but was separated out. It is essentially a 1/3 segment of a hexagonal crystal.

Adapted from: International Union of Crystallography, 2003. Introduction to the crystal class models and from Nelson, 2013

Density vs. Specific Gravity

Density and specific gravity are similar but they are not the same, although they are usually close numerically in value.

"Density (ρ) = mass / volume" and usually expressed in g/cm^3, or something equivalent.

Specific gravity (SG) is a ratio and thus unitless.

Density is usually tested at or near standard temperature and pressure (STP). National Institute of Standards and Technology (NIST) sets that at:

20°C = 293.15 K = 68°F at 1 atmosphere (atm) = 101.325 Kpa = 14.6959 PSI

Specific gravity is a ratio relative to:

1 g = 1 cm^3 = 1 ml of water at 4°C. The exact temperature is 3.98°C but 4°C is close enough.

If your water (deionized) is kept at 4oC, then you can test and calculate SG the same way you test and calculate density.

If you do geotechnical work in the U.S., you'll know ASTM D-792 is the general standard used for density and specific gravity. It uses a different STP, and employs a ridiculously over complicated methodology.

You can loosely convert density to specific gravity by multiplying g/cm^3 by 0.9976.

Optics (for thin sections)

Michel-Lévy (Birefringence) Color Chart

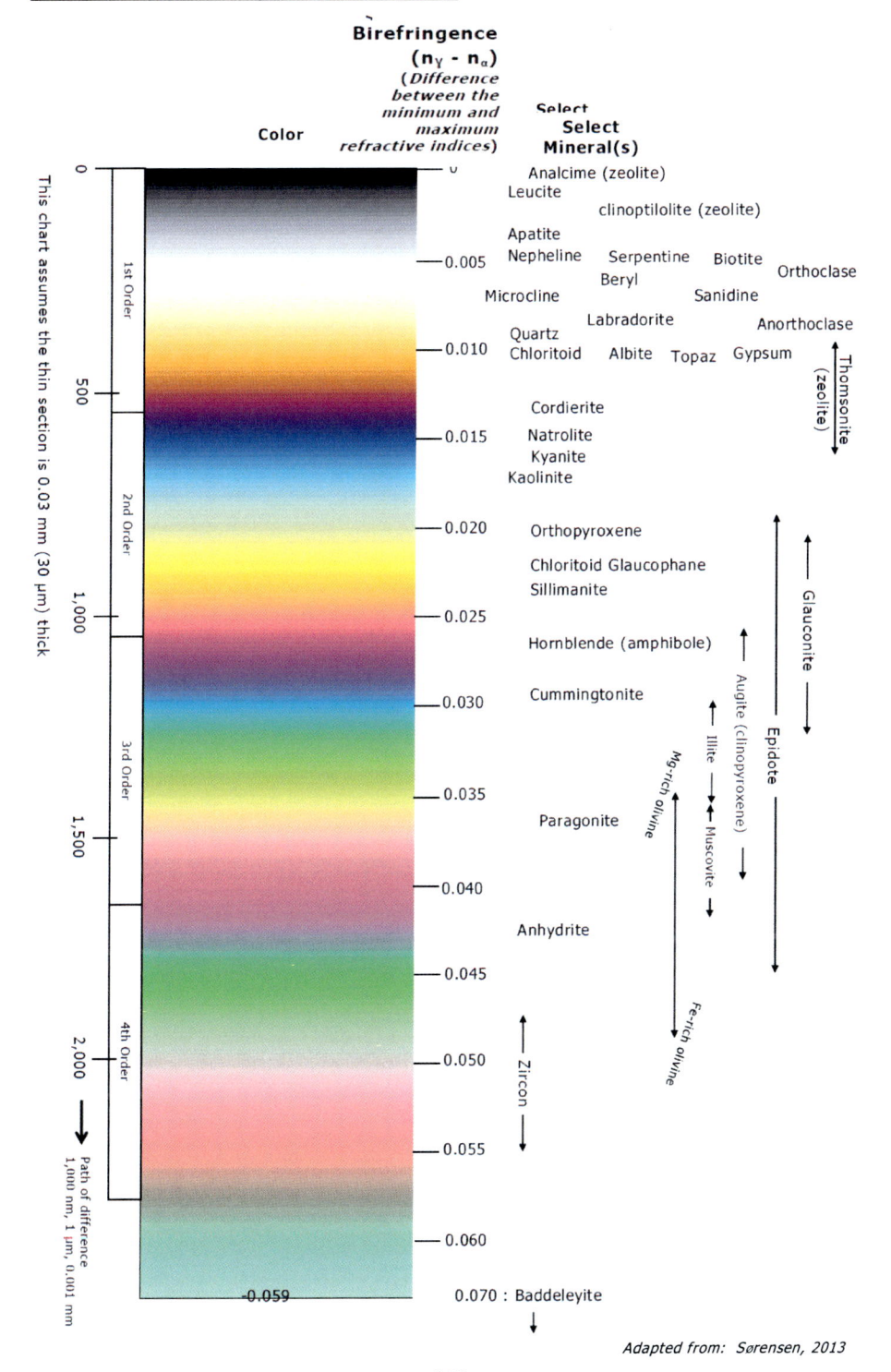

Adapted from: Sørensen, 2013

Birefringence is the double refraction of ordered materials that are transparent to slightly opaque. Minerals will sometimes change to bright colors under polarized light when cut into thin sections. Sunlight and most household light is unpolarized.

Common minerals with no birefringence

Minerals in the isometric (cubic) system have NO birefringence, some common examples are:

- Fluorite
- Galena
- Halite
- Pyrite

Common minerals with high birefringence or negative birefringence

The Michel-Lévy chart only goes down to about 0.050. But many common minerals have a much higher number.

Mineral	Birefringence
Calcite	0.154 to 0.173
Dolomite	0.179 to 0.181
Hematite	0.280
Siderite	0.242
Tourmaline	-0.031

Relative particle sizes for comparison

The following shapes are to scale. I decided to compare a round (circle) object with an angular (hexagonal) object so you can get a better idea of how particles can relate to one another. As you can see, anything smaller than 1 millimeter it is very difficult to decern shape.

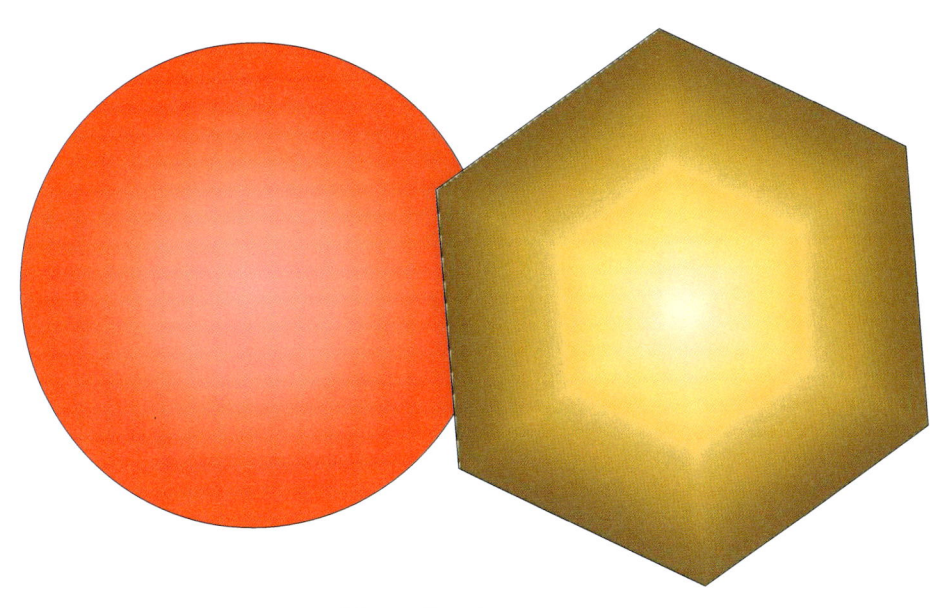

64 mm diameter
(U.S. standard sieve = 2 1/2")

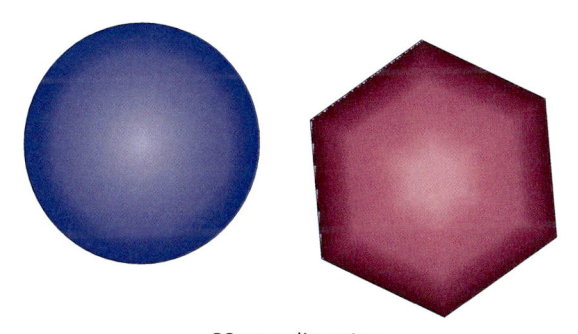

32 mm diameter
(U.S. standard sieve = 1 1/4")

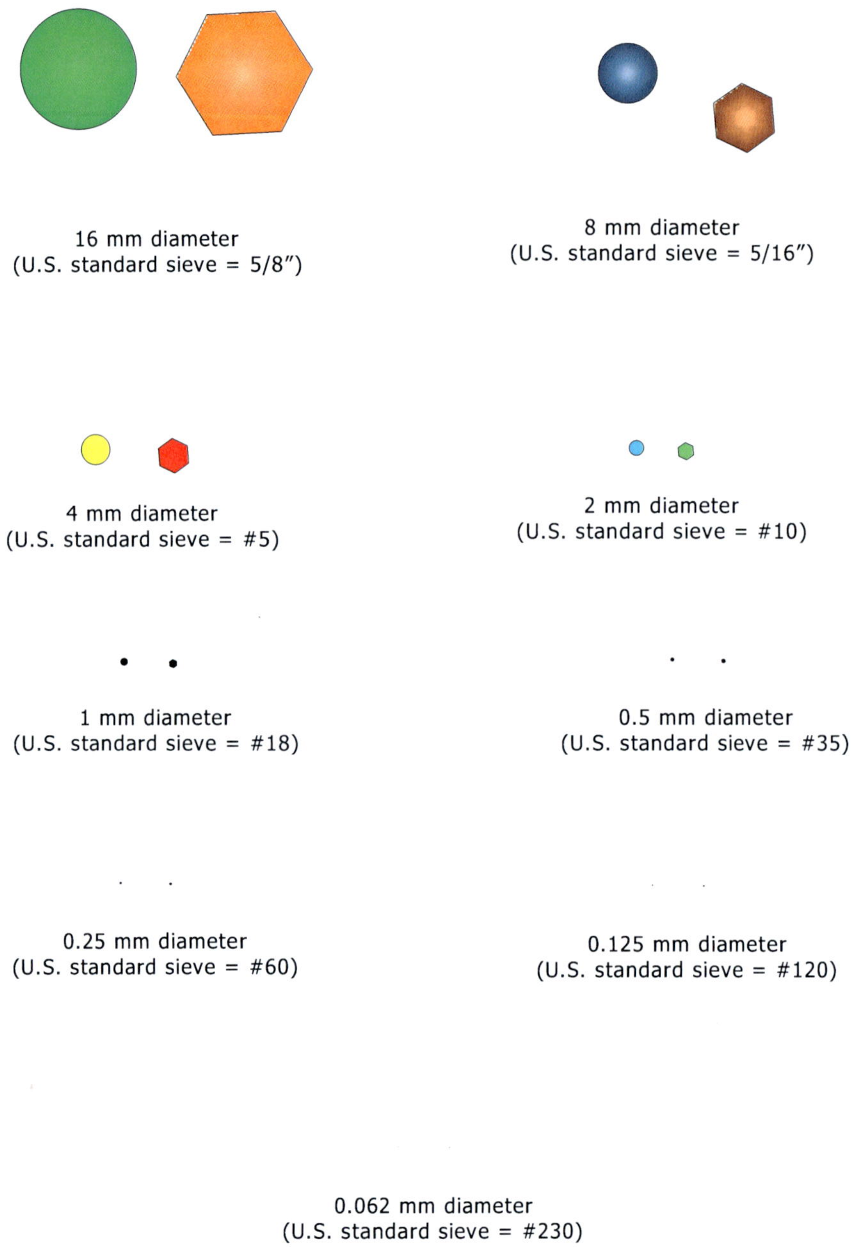

16 mm diameter
(U.S. standard sieve = 5/8")

8 mm diameter
(U.S. standard sieve = 5/16")

4 mm diameter
(U.S. standard sieve = #5)

2 mm diameter
(U.S. standard sieve = #10)

1 mm diameter
(U.S. standard sieve = #18)

0.5 mm diameter
(U.S. standard sieve = #35)

0.25 mm diameter
(U.S. standard sieve = #60)

0.125 mm diameter
(U.S. standard sieve = #120)

0.062 mm diameter
(U.S. standard sieve = #230)

International Mineralogical Association (IMA) abbreviations

Mineral	Abbreviation
Acanthite	Aca
Actinolite	Act
Adularia	Adl
Aegirine	Aeg
Aeschynite-Ce	Aes-Ce
Aeschynite-Y	Aes-Y
Alkali-feldspar	*K-spar* / Afs
Alabandite / alabandine	*Ala*
Albite	Ab
Allanite / orthite subgroup	Aln
Allemontite [1] / Stibarsen	*Alle* / Sbr
Allophane	Alp
Altaite	Alt
Althupite	Ahp
Alunite	Alu
Amakinite	Amk
Amphibolite Group	Amp
Amblygonite	Amy
Analcime / analcite	Anl
Anatese	Ant
Andalusite	And
Andesine [1]	*Ande*
Andradite	Adr
Anglesite	Ang
Ankerite	Ank
Annabergite	Anb
Anorthite	An
Anorthoclase	Ano
Anthophyllite	Ath
Anhydrite	Anh
Antlerite	Atl
Apatite	Ap
Apophyllite	Apo
Aragonite	Arg
Arfvedsonite	Arf
Arsenopyrite	Apy
Atacamite	*Ata*
Augite	Aug
Aurichalcite	Ach
Autunite	Aut
Awaruite	Awr
Axinite	Ax
Azurite	Azu

Barite / Baryite [2]	Brt
Beidellite	Bei
Beryl	Brl
Biotite	Bt
Bismite	Bis
Bismuth	Bi
Boehmite / Böhmite [2]	Boh
Borax	Brx
Bornite	Bn
Boulangerite	Bou
Brannerite	Bnr
Brochantite	Bct
Bronzite [1]	Brz
Brookite	Brk
Brucite	Brc
Bytownite [1]	*Byt*
Calavarite	Clv
Calcite	Cal
Callaghanite	Cgh
Calomel	Clo
Cancrinite	Ccn
Carnallite	Can
Carnotite	Cnt
Celestite / celestine	Clt
Celsian	Cls
Cerussite	Cer
Cervanite	Cvl
Chabazite	Cbz
Chabazite-Ca	Cbz-Ca
Chabazite-K	Cbz-K
Chabazite-Ma	Cbz-Ma
Chabazite-Mg	Cbz-Mg
Chalcocite	Ccn
Chalcophyllite	Chp
Chalcopyrite	Ccp
Chlorite	Chl
Chloritoid	Cld
Chondrodite	Chn
Chromite	Chr
Chromium	Cr
• Chrysoberyl	Cbrl
Chrysocolla	Ccl
Cinnabar	Cin
Clinoclase / Abichite	Cno
Clinoenstatite	Cen

Clinoferrosilite / Clinohypersthene	Cfs
Clinohumite	Chu
Cobaltite	Cbt
Colemanite	Cole
Copper	Cu
Cordierite / dichroite / iolite	Cde
Corundum	Crn
Covellite	Cvl
Cummingtonite	Cum
Cuprite	Cpr
Diamond	Dia
Dickite	Dck
Digenite	Dg
Diopside	Di
Dolomite	Dol
Donpeacorite	Don
Dumortierite	Dum
Edenite	Ed
Enstatite	En
Epidote	Ep
Epsomite	Esm
Esseneite	Ess
Euclase	Ecs
Eulite (*see Ferrosilite*)	N/A
Euxenite	Eux
Fayalite	Fa
Fergusonite-Ce	Fgs-Ce
Fergusonite-Y	Fgs-Y
Ferrosilite	Fs
Fluckite	Fck
Fluorite	Flr
Forsterite	Fo
Franklinite	Frk
Gadolinite / ytterbite [1]	Gad / *Yte*
Gahnite	Ghn
Galaxite	Glx
Galena	Gn
Garnet Group	Grt
Gersdorffite	Gdf
Gibbsite / hydrargillite [1]	Gbs / *Hyd*
Glauberite	Glb
Glaucochroite	Glc
Glauconite	Glt
Glaucophane	Gln
Gmelinite-Ca	Gme-Ca

Gmelinite-K	Gme-K
Gmelinite-Na	Gme-Na
Goethite	Gth
Gold	Au
Graphite	Grt
Grossularite [1] / grossular (gooseberry garnet)	*Gru* / Grs
Gypsum	Gp
Halite (table salt)	Hl
Halloysite	Hly
Hausmannite	Hsm
Hectorite	Htr
Hedenbergite	Hd
Hematite	Hem
Hercynite	Hc
Hessite	Hes
Hornblende minerals	Hbl
Huebnerite / Hübnerite [2]	Hbr
Humite	Hu
Hyalophane [3]	Hyf
Hydrozincite / maronite [1] (zinc bloom)	HZnc / *Mne*
Hypersthene [1] / a ferroan estatite [1]	*Hyt*
Ice	Ice
Idocrase [1] / vesuvianite	*Ido* / Ves
Illite / hydromica [1]	Ilt / *Hda*
Ilmenite	Ilm
Ilvaite / yenite [1]	Ilv / *Yen*
Iodyrite [1] / iodargyrite	*Iod* / Iag
Iridium	Ir
Iron	Fe
Jacobsite	Jcb
Jadeite	Jd
Jamesonite	Jms
Jervisite	Je
Johannsenite	Jhn
kanoite	Knt
Kaoline	Kn
Kaolinite	Kln
Kirkite	Krk
Kerschsteinite [1] / kirschteinite / Monticellite	*Kls* /Klr /Mtc
Kosmochlor	Kos
Kyanite / cyanite [1]	Ky / *Cyn*
Labradorite	Lab
Langbeinite	Lbn
Larsenite	Lsn
Laumonite	Lmt

Lawsonite	Lws
Lazurite	Lzr
Lechatelierite	Lch
Lepidolite	Lpd
Leucite	Lct
Liebenbergite	Lbb
Linnaeite	Lln
Loellingite [1] / Lölingite	Lo / Lö
Maghemite	Mgh
Magnesite	Mgs
Magnetite	Mag
Malachite	Mlc
Manganite	Mnn
Manganosite	Mng
Marcasite	Mrc
Margarite	Mrg
Marialite	Mri
Meionite	Mir
Melonite	Mlt
Metaborite	Mbo
Metacalciouranoite	Mcu
Meyrowitzite	Mey
Microcline	Mcc
Microlite	Mic
Minium	Mnm
Molybdenite	Mol
Monazite group (overall)	Mnz
Monazite-Ce	Mnz-Ce
Monazite-La	Mnz-La
Monazite-Nd	Mnz-Nd
Monazite-Sm	Mnz-Sm
Monticellite (see Kerschsteinite)	
Moolooite	Moo
Mozartite	Moz
Müllerite	Mül
Mullite / porcelainite	Mul
Muscovite	Ms
Nacrite	Ncr
Nagyagite	Nah
Namansilite	Nms
Natalyite	Nta
Natrolite	Ntr
Nchwaningite	Nwg
Nepheline	Nph
Nickel	**Ni**

Nickeline / niccolite	Nc / N/A
Nierite	Nr
Nitre / niter	N/A / Nit
Nontronite	Noo
Noonkanbahite	Non
Nosean / noselite	Nsn / N/A
Odinite	Odn
Ojuelaite	Ojl
Oligoclase [1]	*Olg*
Olivine	Ol
Omphacite	Omp
Opal	Opl
Oppenheimerite	Ohm
Orpiment	Orp
Orthoclase	Orp
Ottrelite	Otr
Oxybismutomicrolite	Obmic
Paakkonenite / Pääkkönenite [2]	Pä
Paarite	Paa
Pargasite [1]	Pge
Palladinite	Pdn
Palladium	**Pd**
Paragonite	Pg
Paulmoorelite	Pmo
Pearceite	Pea
Pectolite	Pct
Pentlandite	Pn
Periclase	Per
Perovskite	Prv
Petalite / castorite [1]	Ptl / *Cst*
Petedunnite	Pdu
Petzite	Ptz
Phenaktite	Phk
Phlogopite	Phl
Phosgenite	Pho
Pigeonite	Pgt
Plagionite	Pgi
Platinum	**Pt**
Polybasite	Plb
Polycrase-Y	Plc-Y
Powellite	Pwl
Prehnite	Prh
Proustite	Prs
Pseudobolenite	Pbol
Pyrargyrite "ruby silver"	Pyg

Mineral	Abbreviation
Pyrite	Py
Pyrochlore	Pcl
Pyrolusite	Pyl
Pyromorphite	Pym
Pyrope	Prp
Pyrophyllite	Prl
Pyroxene	Px
Quartz	Qz / *Qtz*
Realgar "ruby sulfur, ruby arsenic"	Rlg
Rhodochrosite	Rds
Rhodonite	Rdn
Riebeckite	Rbk
Roscoelite	Rcl
Rutile	Rt
Saamite	Saa
Sanidine	Sa
Saponite	Sap
Scolecite	Slc
Scorodite	Scd
Scottyite	Sty
Sedovite	Sdv
Semseyite	Ssy
Serpentine	Srp
Siderite / chalybite	Sd / *Cha*
Sillimanite	Sil
Silver	**Ag**
Smectite	Sme
Smithsonite / calamine [1] / "zinc spar"	Smt / *Cae*
Sodalite	Sdl
Sperrylite	Spy
Spessartine	Sps
Sphaerobertrandite	Sbtd
Sphalerite	Sp
Sphene [2] / titanite	*Sph* / Ttn
Spinel	Spl
Spodumene	Sgg
Staurolite	St
Sternbergite	Srn
Stibnite / antimonite [1] / "kohl"	Sbn / *Atmt*
Stilbite-Ca	Stb-Ca
Stilbite-Na	Stb-Na
Stillwellite-Ce	Swl-Ce
Strontianite	Str
Sulphur	**S**
Sulphur-ß	S-ß

Svanbergite	Svb
Sylvanite	Syv
Sylvite / "potassium salt"	Syl
Talc	Tlc
Tantalite-Fe	Ttl-Fe
Tantalite-Mg	Ttl-Mg
Tantalite-Mn	Ttl-Mn
Tennantite-Cu	Tnt-Cu
Tennantite-Fe	Tnt-Fe
Tennantite-Hg	Tnt-Hg
Tennantite-Zn	Tnt-Zn
Tephroite	Tep
Tetrahedrite-Fe	Ttr-Fe
Tetrahedrite-Hg	Ttr-Hg
Tetrahedrite-Zn	Ttr-Zn
Thalliumpharmacosiderite	Tpsd
Thorianite	Tho
Thorite	Thr
Tin	**Sn**
Titanomagnetite	Tmgh
Topaz	Tpz
Tourmaline	Twn
Tremolite	Tr
Tridymite	Trd
Triphylite	Trp
Trona	Tn
Tungstite	Tgs
Tyuyamunite	Tyu
Uraninite / pitchblende	Urn / *Pbld*
Uranophane	Urp
Uvarovite	Uv
Vanadinite	Vna
Vermiculite	Vrm
Violarite	Vir
Vivianite	Viv
Vulcanite	Vul
Willemite	Wlm
Witherite	Wth
Wolframite	Wol
Wollastonite	Wo
Wulfenite	Wlf
Xenotime-Y	Xtm-Y
Xenotime-Yb	Xtm-Yb
Yvonite	Yv
Zeolite	Zeo

Zincite	Znc
Zircon	Zrn
Zoisite	Zo

Italics = author's designation
[1] = Not recognized by the IMA
[2] = The spelling IMA uses
[3] = Acknowledged but no symbol designated by the IMA
Bold = mineral and an element (e.g. Cu) or elemental compound (e.g. S_8)
" " = common, informal, chemical name

Adapted from: Warr, 2021

Rock abbreviations

Rock	Abbreviation
Amphibolite	Amf
Andesite	Andt
Anorthosite	Ano
Argillite	Arg
Banded iron formation	BIF
Basalt (non vesicular)	Bas
Basalt (vesicular)	BasV
Carbonate	Carb
Carbonatite	Cart
Chert	N/A
Clay	N/A
Clay "stone"	Cls
Coal	N/A
Conglomerate	Cong
Conglomerate (clast supported)	CongC
Conglomerate (matrix supported)	CongM
Diamictite	Dia
Diamictite (>50% quartz)	DiaQ
Diamictite (>50% feldspars)	DiaF
Diamictite (>50% Lithics)	DiaL
Diabase / Dolerite	Db / Dol
Diorite	Dior
Dolostone (pure)	DS
Dolostone (argillaceous)	DSa
Gabbro	Gab
Gneiss	Gns
Granite	Gran
Granodiorite	Grand
Granulite	Granu
Green sandstone	SSg
Harzburgite	Harz
Ignimbrite / welded tuff	Igni / WeTu
Immature quartzite	QtztI
Lamprophyre	Lamp
Latite	Lat
Lherzolite	Lher
Limestone (pure)	LS
Limestone (argillaceous)	Lsa
Loadstone	LdS
Marble	Mar
Marl	N/A
Migmatite	Mig
Monzonite	Monz

Rock	Abbreviation
Mudstone (silt / clay mix)	MuS
Nepheline clinopyroxene / Ijolite	NefC / Ijo
Nepheline monzogabbro / Essexite	NefM / Ess
Norite	Nor
Peridotite	Per
Phenaktite	Fena
Phonolite	Fon
Protoquartzite	QtztP
Pumice	Pum
Pyroxenite (overall)	Pyx
Quartzite	Qtzt
Quartz diorite	DiorQ
Quartz monzonite	MonzQ
Rhyolite (massive)	Rhy
Sand	N/A
Sandstone (arenite)	Ssa
Sandstone (wacke)	SSw
Schist	Sch
Serpentinite	Serpt
Shale	SH
Silt	N/A
Siltstone	SI
Siderite	Sid
Slate	SL
Syenite	Syt
Tonalite	Tont
Trachyte	Tract
Ultramafic (overall)	Umaf
Websterite	Web

There is no rock equivalent to the IMA (p,). So the above abbreviations are by the author. Any overlap between what I have and what others may have created, is purely coincidental.

Molar thermodynamic data for common minerals at standard temperature and pressure, [298.15 K and 10⁵ Pa (1 bar)]

Mineral	Chemical Formula	Crystal System	Molar Mass a.k.a. Formula weight (kg)	Molar Volume (J bar⁻¹)	Specific Heat Capacity $\Delta H°_f$ (kJ)
Akermanite	Ca₂MgSi₂O₇	Tetragonal	0.27264	9.254	-3866.36
Albite (plagioclase)	NaAlSi₃O₈	Triclinic	0.26222	10.006	-3934.56
Almandine	Fe₃Al₂Si₃O₁₂	Isometric	0.49775	11.511	-5263.52
Analcite / Analcime	NaAlSi₂O₆·H₂O	Orthorhombic	0.22016	9.740	-3309.90
Andalusite	Al₂SiO₅	Orthorhombic	0.16205	5.153	-2588.80
Anorthite (plagioclase)	CaAl₂Si₂O₈	Triclinic	0.27821	10.079	-4233.48
Aragonite	CaCO₃	Orthorhombic	0.10009	3.415	-1207.58
Biotite (mica)	K(Mg,Fe)₃(AlSi₃O₁₀)(F,OH)₂	Monoclinic	0.43353	15.060	-2136.24 ?
Brucite	Mg(OH)₂	Trigonal	0.05833	2.463	-924.92
Calcite	CaCO₃	Trigonal	0.10009	3.689	-1207.47
Chloritoid (Fe)	FeAl₂SiO₅(OH)₂	Triclinic (1A polytype) Monoclinic (2A polytype)	0.25191	6.980	-3215.38
Chloritoid (Mg)	MgAl₂SiO₅(OH)₂	Triclinic (1A polytype) Monoclinic (2A polytype)	0.22037	6.875	-3551.42
Clinopyroxene (Ca-Al)	CaAl₂SiO₆	Monoclinic	0.21813	6.356	-3307.03
Coesite	SiO₂	Monoclinic	0.06009	2.064	-905.47
Corundum	Al₂O₃	Trigonal	0.10196	2.558	-1675.25
Daphnite (Mg variety of chamosite)	Fe₅Al₂Si₃O₁₀(OH)₄	Monoclinic	0.64548	21.340	-7134.85
Diamond	C	Isometric	0.01201	0.342	2.07
Diopside	CaMg(SiO₃)₂	Monoclinic	0.21656	6.619	-3202.76
Dolomite	CaMg(CO₃)₂	Trigonal	0.18441	6.434	-2324.43
Enstatite	Mg₂(SiO₃)₂	Orthorhombic	0.20079	6.262	-3090.47
Epidote	Ca₂FeAl₂Si₃O₁₂(OH)	Monoclinic	0.65175	13.910	-6463.21
Fayalite	Fe₂SiO₄	Orthorhombic	0.20378	4.631	-1478.15
Ferrosilite	Fe₂(SiO₃)₂	Orthorhombic	0.26386	6.592	-2388.63
Forsterite	Mg₂SiO₄	Orthorhombic	0.14071	4.366	-2172.20
Glaucophane	Na₂Mg₃Al₂Si₈O₂₂(OH)₂	Monoclinic	0.78355	26.050	-11969.47
Graphite	C	Hexagonal	0.01201	0.530	0
Grossular (garnet)	Ca₃Al₂Si₃O₁₂	Isometric	0.450454	12.535	-6644.15

284

Mineral	Chemical Formula	Crystal System	Molar Mass a.k.a. Formula weight (kg)	Molar Volume (J bar-1)	Specific Heat Capacity $\Delta H°_f$ (kJ)
Hedenbergite	CaFe(SiO$_3$)$_2$	Monoclinic	0.24811	6.795	-2844.16
Hematite	Fe$_2$O$_3$	Trigonal	0.15969	3.027	-825.71
Ilmenite	FeTiO$_3$	Trigonal	0.15175	3.169	-1231.30
Jadeite	NaAl(SiO$_3$)$_2$	Monoclinic	0.20214	6.040	-3027.85
Kaolinite	Al$_2$Si$_2$O$_5$(OH)$_4$	Triclinic	0.25816	9.934	-4122.18
Kyanite	Al$_2$SiO$_5$	Triclinic	0.16205	4.414	-2593.11
Labradorite (plagioclase)	(Ca,Na)(Al,Si)$_4$O$_8$	Triclinic	0.27181	10.372	-3560.17 ?
Laumontite (zeolite)	CaAl$_2$Si$_4$O$_{12}$·4H$_2$O	Monoclinic	0.47044	20.370	-7268.47
Lawsonite	CaAl$_2$Si$_2$O$_7$(OH)$_2$·H$_2$O	Orthorhombic	0.31420	10.132	-4869.14
Leucite	KAlSi$_2$O$_6$	Tetragonal	0.21825	8.828	-3029.16
Magnesite	MgCO$_3$	Trigonal	0.084321	2.803	-1111.36
Magnetite	Fe$_3$O$_4$	Isometric	0.231539	4.452	-1115.51
Microcline (K-spar)	KAlSi$_3$O$_8$	Triclinic	0.27834	10.892	-3975.11
Monticellite	CaMgSiO$_4$	Orthorhombic	0.15648	5.148	-2253.05
Muscovite (mica)	KAl$_2$(AlSi$_3$O$_{10}$)(OH)$_2$	Monoclinic	0.39831	14.083	-5984.18
Nepheline (feldspathoid)	NaAlSiO$_4$	Hexagonal	0.14523	5.419	-2095.08
Orthoclase (K-spar)	KAlSi$_3$O$_8$	Monoclinic	0.27833	2.803	-3970.04
Paragonite (mica)	NaAl$_2$(AlSi$_3$O$_{10}$)(OH)$_2$	Monoclinic	0.382201	13.211	-5946.34
Prehnite	Ca$_2$Al(AlSi$_3$O$_{10}$)(OH)$_2$	Orthorhombic	0.41239	14.026	-6203.18
Quartz	SiO$_2$	Hexagonal	0.06009	2.269	-910.83
Rutile	TiO$_2$	Tetragonal	0.07990	1.882	-944.18
Sanidine	KAlSi$_3$O$_8$	Monoclinic	0.27734	10.900	-3964.96
Sillimanite	Al$_2$SiO$_5$	Orthorhombic	0.16205	4.986	-2585.68
Spinel	MgAl$_2$O$_4$	Isometric	0.14227	3.978	-2300.72
Staurolite (Fe)	FeAl$_{18}$Si$_{7.5}$O$_{48}$H$_4$	Monoclinic	1.69170	44.880	-23753.93
Talc	Mg$_3$Si$_4$O$_{10}$(OH)$_2$	Triclinic	0.37929	13.625	-5897.1
Ulvöspinel	Fe$_2$TiO$_4$	Isometric	0.22359	4.682	-1497.49
Wollastonite	CaSiO$_3$	Triclinic Monoclinic (polytype)	0.116164	3.993	-1634.06

? = author's calculated estimates

Adapted from: Philpotts and Ague, 2009

BLANK NOTES:

PHOTOS OF SELECT MINERALS

Amphiboles

Cummingtonite
Locality: Lawrence County, South Dakota
Photo by Dave Dyet

Hornblende (dark mineral)
Locality: Buskerud, Norway
Photo by Rob Lavinsky

Glaucophane (blue)
Locality: Brittany, France
Photo by Didier Descouens

Pargasite (dark green)
Locality: Ganesh, Pakistan
Photo by Rob Lavinsky

Olivines

Olivine (Green)
Locality: ?
Photo by Steven Baumann

Tephroite
Locality: Tochigi-ken Honshu, Japan
Photo by Dave Dyet

Photos above have had their backgrounds digitally removed. Photos under Creative Commons Attribution-ShareAlike 4.0 International License : https://en.wikipedia.org/wiki/Creative_Commons, except the photos by the author Steven Baumann

Pyroxenes

Clinopyroxene: Diopside
Locality: Khugiani, Nangarhar (Ningarhar) Province, Afghanistan
Photo by Rob Lavinsky

Clinopyroxene: Aegirine (black)
Locality: Mt. Malosa, Zomba, Malawi
Photo by Ivar Leidus

Orthopyroxene: Enstatite
Locality: Bare Hills Maryland
Photo by Rob Lavinsky

Clinopyroxene: Spodumene
Locality: Huntington Massachusetts
Photo by Rob Lavinsky

Tectosilicates
Feldspars

Plagioclase: Augite
Locality: Piedmont, Italy
Photo by Rob Lavinsky

Plagioclase: Labradorite
Locality: Madagascar
Photo by Parent Géry

Photos above have had their backgrounds digitally removed. Photos under Creative Commons Attribution-ShareAlike 4.0 International License : https://en.wikipedia.org/wiki/Creative_Commons, except the photos by the author Steven Baumann

Plagioclase: Anorthite
Locality: Somma-Vesuvius Complex
Photo by Rob Lavinsky

Plagioclase: Oligoclase
Locality: Sonora, Mexico
Photo by Rob Lavinsky

K-spar: Sanidine
Locality: Auvergne, France
Photo by Didier Descouens

K-spar: Orthoclase
Locality: Minas Gerias, Brazil
Photo by Didier Descouens

Feldspathoids (Foids)

Nepheline
Locality: ?
Photo by Andrew Silver

Sodalite
Locality: Minas Gerias, Brazil
Photo by John Alen Elson

Leucite
Locality: Albano Hills Italy
Photo by Dave Dyet

Quartz

Agate
Locality: Duluth, Minnesota
Photo by Lech Dardki

Quartz
Locality: Baraboo, Wisconsin
Photo by Steven Baumann

Photos above have had their backgrounds digitally removed. Photos under Creative Commons Attribution-ShareAlike 4.0 International License : https://en.wikipedia.org/wiki/Creative_Commons, except the photos by the author Steven Baumann

Zeolites

Thomsonite (pink)
Locality: Mumbi, India
Photo by Rob Lavinsky

Laumontite
Locality: Keweenaw County, Michigan
Photo by Steven Baumann

Photos above have had their backgrounds digitally removed. Photos under Creative Commons Attribution-ShareAlike 4.0 International License : https://en.wikipedia.org/wiki/Creative_Commons, except the photos by the author Steven Baumann

Unit Conversions

abbreviations

Unit	System
ac = acre	imperial
ch = chain	imperial
cm = centimeter	metric
fur = furlong	imperial
ft = foot	imperial
g = gram	metric
gal = gallon	US customary
gi = gill	US customary
gr = grain	N/A
in = inch	imperial
Km = kilometer	metric
l = liter	metric
lb = pound	avoirdupois
li = link	imperial
Ls = light second	SI unit

Unit	System
m = meter	metric
mi = mile	imperial
ml = milliliter	imperial
mm = millimeter	metric
oz(f) = ounce (fluid)	US customary
oz(s) = ounce (solid)	avoirdupois
pt = pint	imperial
qt = quart	US customary
qtr = quarter	imperial
rd = rod	imperial
s = second	SI unit
st = stone	stone
ton = tonne	metric
ts = ton short	imperial
yd = yard	imperial

Area

1 in^2 = 6.4516 cm^2
1 ft^2 = 144 in^2 = 0.092903 m^2
1 yd^2 = 9 ft^2 = 0.836127 m^2
1 m^2 = 10.7639 ft^2 = 1.19599 yd^2
1 km^2 = 0.386102 mi^2 = 247.105 ac
1 acre = 160 rd^2 = 43,560 ft^2 = 4,046.8564 m^2 = 4,840 yd^2
10 ac = 1 fur^2

Length

1 m = 1/299,792,458 Ls = 3.3335641x10^{-9} Ls
1 m = 100 cm = 1,000 mm = 3.28084 ft = 39.37008 in = 1.09361 yd
7.92 in = 20.1168 cm (exact) = 1 li
100 li = 1 ch = 66 ft = 22 yd = 4 rd = 20.1168 m (exact)
1 rd = 0.25 ch = 16.5 ft (exact) = 5.5 yd = 5.0292 m
1 fur = 10 ch = 660 ft = 220 yd = 40 rd
1 mi = 8 fur =20 ch = 5,280 ft = 1,760 yd = 320 rd
0.1 ft = 1.2 in = 3.048 cm = 30.48 mm
0.01 ft = 0.12 in = 0.3048 cm—3.048 mm
1 metric chain = 20 m (exact) = 65.6168 ft (used in India)

Mass

1 grain = 64.79891 mg (exact) = 0.0028571 oz(s)
1 lb =16 oz(s) = 0.453592 kg
1 kg = 1,000 g = 2.20462 lb = 15,432 gr
1 qtr = 28 lb = 11.339809 kg
1 st = 14 lb = 6.35029 kg
2,000 lb = 1 ts = 0.907 tonne = 907.1847 kg

Volume

1 l = 1,000 ml = 0.264172 gal = 1.05669 qt = 8.45351 gi
1 gal = 4 qt = 3.78541 l = 8 pt = 32 gi
1 in^3 = 16.3871 cm^3
1 ft^3 = 1,728 in^3 (exact) = 28,316.8 cm^3
1 yd^3 = 27 ft^3 = 0.764555 m^3
1 cm^3 = 1 ml = 0.0610237 in^3
1 m^3 = 1.30795 yd^3 = 35.3147 ft^3

REFERENCES

6.4.7: Pyroxenes - Geosciences LibreTexts. Downloaded September 20, 2023. under a CC BY-NC-SA 4.0 license and was authored, remixed, and/or curated by Dexter Perkins via source content that was edited to the style and standards of the LibreTexts platform.

Allaby, M., 2008: A dictionary of earth sciences. Oxford University Press, 3rd Edition, ISBN 9780199211944

Anthony, J.W., Bideaux, R.A., Bladh, K.W., and Nichols, M.C. (editors): Handbook of Mineralogy, Mineralogical Society of America, Chantilly, VA 20151-1110, USA., www.handbookofmineralogy.org/

Ashwall, L.D., Morrison, D.A., Phinney, W.C., and Wood, J., 1983: Origin of Archean anorthosites: Evidence from the Bad Vermilion Lake anorthosite complex, Ontario, Contributions to Minerology and Petrology, v.82, p.259-273

Ashwall, L.D., 2010: The temporality of anorthosites, Canadian Mineralogist, v.48, p.711-728

ASTM 2487: Classification of soils for engineering purposes (Unified soil classification system)

Baumann, S.D.J, Cory, A.B., Krol, M.M., Piispa, E.J., 2015: Precambrian geologic events in the Mid-continent of North America, Midwest Institute of Geosciences and Engineering, Publication G-012011-1G

Baumann, S.D.J. and Michaels, E., 2011: Classification of conglomerates and sandstones with charts. Midwest Institute of Geosciences and Engineering, Publication G-022011-1A

Baumann, S.D.J., 2016: Archean outcrop, 1.85 kilometers west of the Green Lakes Trans-Canada 17, Ontario Canada. Midwest Institute of Geosciences and Engineering, publication G-082016-5A

Baumann, S.D.J., 2019: Midwest geologist's reference book (Deluxe edition). Midwest Institute of Geosciences and Engineering, ISBN 9781795849616

Baumann, S.D.J., 2019: Lab reference book; classifying igneous rocks (visual methods). Midwest Institute of Geosciences and Engineering, ISBN 9781797802497

Baumann, S.D.J., 2019: Lab reference book; classifying sedimentary rocks (visual methods). Midwest Institute of Geosciences and Engineering, ISBN 9781799025801

Boggs Jr., S., 2009: Petrology of sedimentary rocks. 2nd edition, Cambridge University Press, ISBN: 9780511719332

Bouma, A.H., Ravenne, C., 2004: The Bouma sequence (1962) and the resurgence of geologic interest in the French Maritime Alps (1980s): The influence of Gres d'Annot in developing the turbidite systems. Geological Society of London, Special Publications, v.221, p. 27-38

Bowen, N.L., 1922: The reaction principle in petrogenesis, Journal of Geology, v.XXX, no.3, p.177-198

Bucher, K., and Grapes, R., 2011: Petrogenesis of metamorphic rocks. Springer Heidelberg Dordrecht London New York, 1st Edition, ISBN 9783540741688

Cain, J.A. and Beckman, W.A., 1964: Preliminary report on the Precambrian geology of the Athelstane area, northeast Wisconsin, Ohio Journal of Science, v.64 no.1 p,57-60

Campbell, C.V., 1967: Lamina, laminaset, bed and bedset. Sedimentology, v.8, p. 7-26

Casshyap, S.M., 1968: Huronian stratigraphy and paleocurrent analysis in the Espanola-Willisville area, Sudbury District, Ontario Canada. Sedimentary Petrology, vol. 38, no. 3, pp. 920-942

Chandler, B.M,P., Boston, C.M., Likas, S., Lovel, H., 2021: Re-interpretation of 'hummocky moraine' in the Gaick, Scotland, as erosional remnants: Implications for palaeoglacier dynamics. Proceedings of the Geologists' Association 132 (2021) 506–524

Dunham, R. J., 1962: Classification of carbonate rocks according to depositional texture. In: Ham, W. E. (ed.), Classification of carbonate rocks: American Association of Petroleum Geologists Memoir, p. 108-121

Fabries, J. et al., 1989: Nomenclature of pyroxenes, Mineralogical Journal, v.14, no.5, p.198-221

Fedele, L., Lustrino, M., Melluso, L., Morra, V., Zanetti, A., and Vannucci, R., 2015: Trace-element partitioning between plagioclase, alaki feldspar, T

magnetite, biotite, apatite, and evolved potassic liquids from Campi Flegrei (southern Italy), American Mineralogist, v.100, p.233-249

Flint, R.F., Sanders, J.E., and Rodgers, J., 1960: Diamictite a substitute term for symmictite. Geological Society of America, Bulletin v.71, p. 1809-1810

Fluorescent colors on p.26 and p.29-32 of this book were taken from: fluomin.org/uk/

Foley, D., 2009: Mineralogy and metamorphism of the stromatolite-bearing layers of the Biwabik Iron Formation. B.A. thesis (Geology), Gustavus Adolphus College

Folk, R.L., 1954: The distinction between grain size and mineral composition in sedimentary rock 606 nomenclature. Geology, v.62, p. 344-359.

Folk, R.L., 1980: Petrology of sedimentary rocks. Hemphill Publishing Company, ISBN: 0914696149

Franklin mineral museum, fluorescent mineral list: www.franklinmineralmuseum.org/fluorescent-minerals/#:~:text=X%20Y%20Z-,A,V%5D%20%7BFO%7D. Retrieved October 5, 1024

Fritz, W.J., 1993: Tuff or sandstone?--Usage of size and compositional terms in the classification of pyroclastic rocks, IAVCEI Commission on Volcanic Sediments, Newsletter, no.8, p1-2

Frost, C.D., Frost, B.R., Kirkwood, R., Chamberlain, K.R., 2006, The tonalite-trondhjemite-granodiorite (TTG) to granodiorite-granite (GG) in the late Archean plutonic rocks of the central Wyoming province, Canadian Journal of Earth Science, v.43, p.1419-1444

Fuerstenau, D.W., Hansen, J.S., and Diao, J., 1994: A simple procedure for the determination of pyritic sulfur content in coal. Fuel, v.73, no.1, p. 123-127

Gill, R., 2010: Igneous rocks and processes, Wiley-Blackwell publishing, ISBN 13: 978-1-4443-3065-6

Gillespie, M.R. and Styles, M.T., 1999: BGS Rock classification scheme, Classification of igneous rocks, Research report number RR 99-06, Volume 1

Gordon, L.M., 2015: Atom probe tomography of feldspars and aluminosilicate glasses, Microscopy and microanalysis, no.0427, p.853-854

Greensmith, J.T., 1989: Petrology of the sedimentary rocks. 7th edition, Unwin Hyman Ltd., ISBN: 9789401196406

Hansel, A.K. and Johnson, W.H., 1996: Wedron and Mason Groups: Lithostratigraphic reclassification of deposits of the Wisconsin episode, Lake Michigan lobe area. Illinois State Geological Survey, Bulletin 104

Higgins, M.W., 1971: Cataclastic rocks. United States Geological Survey, Professional Paper 687

Hoffman, P.F. and Randalli, G., 1988: Archean ocean flake tectonics, Geophysical Research Letters, v.15, no.10, p.1077-1080

Huang, H., Polat, A., Fryer, B.J., 2013: Origin of Archean tonalite-trondhjemite-granodiorite (TTG) suites and granites in the Fiskenaesset region, southern west Greenland: implications for continental growth, Gowganda Research, v.23, p.452-470

Kamvisis, I., Phani, P.R.C., 2022. The Lamprophyre Clan Revisited. J Geol Soc India 98, 1205–1209 (2022). https://doi.org/10.1007/s12594-022-2153-4

Kurniawan, A., Mc. Kenzie, J., and Putri, J.A., 2009: General dictionary of geology. Environmental Geographic Student Association, 1st Edition.

LeBas, M.J. and Streckeisen, A.L., 1991: The IUGS systematics of igneous rocks, Journal of the Geological Society of London, v.148, p.825-833

Leith, C.K., 1905: Rock cleavage. United States Geological Survey, Bulletin 239

LeMaitre, R.W. (editor), Streckeisen, A., Zanettin, B., LeBas, M.J., Bonin, B., Bateman, P., Bellieni, G., Dudek, A., Efremova, S., Keller, J., Lameyre, J., Sabine, P.A., Schmind, R., Sorensen, H., Woolley, A.R., 2002: Igneous rocks, a classification of terms, recommendations of the International Union of Geological Sciences subcommission on the systematics if igneous rocks, Cambridge University Press in the UK, published in the US by Cambridge University Press, New York, ISBN-13: 978-0-521-66215-4, ISBN 10: 0-521-66215-X

Licker, MD. et al., 2003: Dictionary of Geology and Mineralogy, 2nd edition, McGraw-Hill, ISBN 0-07-141044-9

Lightfoot, P.C., 2017. Chapter 3 - Petrology and geochemistry of the Sudbury igneous complex, Nickel Sulfide Ores and Impact Melts. Elsevier, Pages 190-295.e1, ISBN 9780128040508, doi.org/10.1016/B978-0-12-804050-8.00003-1.

Lindsey, D.A., 2007: Precambrian time—the story of early Earth. United States Geological Survey, Fact Sheet 2007-3004

Mahapatra, D. 2016: A review on steam coal analysis-sulfur. Research in Science and Technology, engineering and mathematics, ISSN: 23283580

McBride, E.F., 1963: A classification of common sandstones. Sedimentary Research, v.33, no.3, p. 664-669

Mills, S.J., Hatert, F., Nickel, E.H., and Ferraris, G., 2009: The standardization of mineral group hierarchies: application to recent nomenclature proposals. European Journal of Mineralogy, v. 21, pp. 1073-1080, DOI: 10.1127/0935-1221/2009/0021-1994

Nelson, S.A., 2013: External symmetry of crystals, 32 crystal classes, Tulane University, EENS 2110, mineralogy

Nelson, S.A., 2017: Metamorphism and metamorphic rocks. Lecture notes, Physical geology class EENS1110, Tulane University

Nelson, S.A., 2018: Types of metamorphism. Lecture notes, Petrology class EENS2120, Tulane University

Nyfeler, D. et al., 1995: chwaningite, $Mn^{~+}SiO_3(OH)_2 \cdot H_2O$, a new pyroxene-related chain silicate from the N'chwaning mine, Kalahari manganese field, South Africa, American Mineralogist, v.80, p.377-386

Pawley, M., Reid, A., Dutch, R., and Preiss, W., 2013: A user's guide to migmatites. Government of South Australia, Department for Manufacturing, Innovation, Trade, Resources and Energy

Philpotts, A.R. and Ague, J.J., 2009. Principles of igneous and metamorphic petrology, 2nd edition, Cambridge University Press, ISBN: 978-0-521-88006-0

Rich, J.L., 1951: THREE CRITICAL ENVIRONMENTS OF DEPOSITION, AND CRITERIA FOR RECOGNITION OF ROCKS DEPOSITED IN EACH OF THEM, GSA Bulletin (1951) 62 (1): 1–20

Robertson, S., 1999: BGS rock classification scheme, classification of metamorphic rocks. British Geological Survey Report, RR 99—02, vol. 2

Soloviev, S.G., and Kryazhev, S., 2017: Geology, mineralization, and fluid inclusion characteristics of the Chorukh-Dairon W-Mo-Cu skarn deposit in the middle Tien Shan, northern Tajikistan. Ore Geology Reviews, vol. 80, pp. 79-102

SØrensen, B.E., 2013: A revised Michel-Lévy interference colour chart based on first-principles calculations. European journal of mineralogy, DOI: 10.1127/0935-1221/2013/0025-2252

Strunz, H. and Nickel, E.H., 2001: Strunz mineralogical tables. Schweizerbart, Stuttgart, 870 p., ISBN 9783510651887

Tucker, M.E., 2003: Sedimentary rocks in the field. 3rd edition, John Wiley and Sons Ltd., ISBN 0470851236

Tucker, M.E., 2006.Sedimentary petrology. 3rd edition, Blackwell Publishing, ISBN-13: 9780632057351

United States Geological Survey, 2015: www.usgs.gov: What is the difference between a rock and a mineral?

Ure, A., 1840: Dictionary of arts, manufactures, and mines. Longmans Green and Co., p. 619

Ure, A. and Rudler, F.W., 1878: Dictionary of arts, manufactures, and mines. Volume II, Longmans Green and Co. p. 736-737

VanSchmus, W.R. and Hinze, W.J., 1985: The Midcontinent rift system, Annual Review, Earth Planet, Science, V.13, p.345-383

Vernon, R.H. and Clarke, G.L., 2008: Principles of metamorphic petrology. Cambridge University Press, ISBN 9780521871785

Warr, L.N., 2021: IMA–CNMNC approved mineral symbols, Mineralogical Magazine (2021), 85, 291–320, doi:10.1180/mgm.2021.43

Wentworth, C.K., 1922: A scale of grade and class terms for clastic rocks. Geology, v.30, no.5, p. 377-392

Wimmenauer, W., and Bryhni, I., 2007: Migmatites and related rocks. A proposal on behalf of the IUGS subcommission on the systematics of metamorphic rocks, web version 01.02.07

Winter, J.D., 2001: Igneous and metamorphic petrology. Prentice Hall Upper Saddle, 1st Edition, ISBN 0132403420

Index of Figures

A More Realistic Cross Section of the interior of the Earth   241
Accepted Pyroxene Chemical Subdivisions   89-91
Adapted Wentworth Scale Chart for Crystalline Size   117
Aphanitic (volcanic) Rock Identification Flowchart Based on Color   60
Basic Properties of Select Rocks and Minerals / Mineraloids   252-265
Bedding Planes   178
Bedding Thickness Terms   183
Bowen's Reaction Series   26-27
Bowen's Reaction Tree   28
Carbonate ternary plot based off mineralogy   218
Change in Volume of Select Clay Mineral Groups   207
Classification of Diamicton   209
Classification of Lamprophyres 39-40
Clinopyroxene (Cpx): Monoclinic   88
Color of Lithological Units   246-247
Common Igneous Minerals   21-22
Common Metamorphic Videos 122-127
Common Sedimentary Minerals   186-188
Compositional Phase Diagrams (Feldspars)   42
Compositional. Textural, Rock Name Block Diagram   79
Conglomerates with Matrix (Ternary Plot)   193
Conglomerites with Matrix (Ternary Plot)   145
Cross Bed Patterns   182
Cross Sectional View of a Typical Bouma Cycle   202
Crystal Systems   266-267
Crystalline Carbonates   216
Crystalloblastic Series   115
Euhedral, Subhedral, and Anhedral   114
Foliations and Lineations to Scale   131
Flow Chart for Testing Carbonates   215
Geothermal Gradient   238
Gold vs. Pyrite and Chalcopyrite   96-97
Grain Angularity   170
Grain Distribution   176
Grain Percentages Based Off Volume   248-250
Grain Shape   175

Grain Sorting   166

Graphic Depiction of Cleavage   132

Igneous Rock Samples Under High Magnification   102-104

Intergrain Relationships   167-168

Internal Mechanical Layers of the Earth and Their Percent Mass   239

Internal Mechanical Layers of the Earth and Their Percent Volume   240

International Mineralogical Association (IMA) abbreviations   273-281

Ironite (Ternary Plot)   148

IUGS QAPF Plot for Aphanitic (Volcanic) Rocks   65

IUGS QAPF Plot for Phaneritic (Plutonic) Rocks   33

Lithological Carbonate Symbols   219-220

Lithological Chertstone, BIF, Evaporite, and Coal Symbols   229

Lithological Conglomerate Symbols   195

Lithological Mudstone Symbols   210

Lithological Sandstone Symbols   203

Marble Ternary Plot Based off Mineralogy   149

Maturity/Texture 3D Diagram (Sedimentary)   199

Metamorphic Facies Temperature and Pressure Diagram   120

Metamorphic Grade Temperature and Pressure Diagram   119

Metamorphic Rock Types   113

Michel-Levy Birefringence Color Chart   269

Midwest Basins   222

Metamorphic Rocks Under High Magnification   156-159

Modified Wentworth Size Scale Chart for Igneous Rocks   20

Modified Wentworth Size Scale Chart for Igneous Rocks (With Lamprophyre and Komatiite)   41

Moh's Hardness   21

Molar Thermodynamic Data for Common Minerals at Standard Temperature and Pressure, [298.15 K and 105 Pa (1 bar)]   284-285

Non Crystalline Grains (Carbonates)   217

Other Coal Components   228

Orthopyroxene (Opx): Orthorhombic   87

Particle Size Comparisons   174

Phaneritic Mafic Rock Chart   36-37

Phaneritic Ultramafic Rock Chart   38

Photos of Select Minerals   305-308

Positive Skew Indicates Coarse Grains Exhibit Better Sorting Than Fine   177

Pseudo-cubic (pseudo-isometric) trapezohedron crystal   51

Pyroclastic Rock Classification Diagram   76

Pyroclastic Sediment Classification Diagram   77
Pyroxenite (Ternary Plot)   92-93
QFL Arenites (Ternary Plot)   197
QFL Wacke (Ternary Plot)   198
Quick Flowchart for Choosing Plagioclase or Alkali Feldspar   44
Relative Amounts of Minerals in Felsic v. Intermediate v. Mafic v. Ultramafic Rocks   80
Relative Hardness Chart   185
Relative Particle Sizes for Comparison   271-272
Rock Abbreviations   282-283
Sandite Classification Chart (Ternary Plot)   141
Sedimentary Rock Types Based off Traditional Grouping   189
Sedimentary Rocks Under High Magnification   230-232
Silicate/Oxide/Carbonate Iron [SOC(Fe)] Diagram   224
Simplified Classification Comparison Chart (Igneous Rocks)   82
Simplified Flow Chart for Classifying Igneous Rocks   31
Stability Fields of Minerals in BIF   147
Striations on Plagioclase vs. Microcline   43
Types and Causes of Metamorphism   118
Types of Coal   227
Types of Cleavage   132
Type of Mudstones   205
Undersaturated vs. Saturated   25
Wentworth Scale   173
Worksheet for QAPF Plot   101

Index by Subject

A

Albite 22, 26, 28 **45**, 50, 51, 85, 88, 121, 122

Alkali feldspar 17, 22, 23, 33, 24, 40, **42**-45, 49-51, 53, 56, 57, 60, 62, 65, 66-69, 73, 81, 99, 100, 196

Alkali feldspar syenite **33**

Amphibole / amphibolite 52, 79, 85-87, 115, 119, 120, 122-124, 126, 130, 137, 146, 282, 305

Andesine 22, 26. 28, 42, **46**, 53

Andesite 46, 59, 60, 62-64, 82, 88

Anhedral 48, 64, 88, 102, 114

Ankerite 50, 51, 122, 186, **225**, 273

Anorthite 22, 26-28, 30, 35, **48**, 53, 70, 154, 307

Anorthosite 16, 25, 27, 33, **70**, 81

Anorthoclase 22, 42

Apatite 21, 22, 52, 186

Aphanitic 20, 25, 31, 34, 58, **59**, 60, 64-66, 78, 98, 104

Argillite 282

Arterite Migmatite 151, 152

Ash 75-77, 130, 135-137, 228

Augite 22, 84, 86, **88**, 104. 122, 126, 282, 306

B

Banded iron formation (BIF) 113, 124, 146-148, 154, 162, 164, 184, 187, 221, **223**-226, 229, 232, 246, 282

Basalt 54, 59, 60, 63-**65**, , 74, 75, 79, 82, 86, 143, 155

Bentonite 206, 207

Beryl 22, 88, 275

Birefringence 269, 270

Bio-carbonate 169

Biotite 21, 22, 26, 28, 46, 49, 52, 56, 57, 72, 78-80, 82, 85, 99, 100, 102, 121, 123, 127, 159, 186

Bismite 275

Bismuth 275

Blueschist 120, 124, 125

Boehmite / Bömite 275

Borax 275

Bouma cycle sequence **202**

Boundstone 217

Bowen's reaction series 26-30, 42, 72. 73, 81, 83, 115, 124, 153

Bowen's reaction tree 28

Breccia 75-77, 95, 129, 133, 162,191-192

Bronzite 275

Bytownite 22, 26, 28, **47**, 70, 275

C

Calavarite 275

Calcite 31, 22, 50, 51, 63, 64, 78, 112, 115, 123, 165, 186, 195, 212, 216, 236

Carbon 124, 184

Carbonate 16, 17, 31, 64, 87, 122, 139, 148, 154, 155, 157, 162, 163, 165, 186, 189, 191, 196, **212**, 213, 215, 216-225, 232

Celestite / celestine 275

Chalcocite 275

Chalcopyrite 94, **96**, 97, 104,

Chlorastrolite 22, 23

Chlorite 22, 63, 115, 121, 123, 156, 159, 206, 275

Chromite 275

Chromium 275

Clay 19, 20, 51, 117, 146, 169, **173**, 200, 201, 204, 205-207-210, 212, 217, 227, 282

Clinopyroxene 86, 87, **88**, 121-123, 125, 126, 306

Color 16, 35, 44,-52, 57, 59, 60, 64, 70, 78, 83-86, 94, 96, 98, 100, 103, 104, 122-126, 133, 156, 157, 184, 194, 196, 201, 206, 222, 230-232, 241, **245**, 246, 270

Conglomerate 113, 133, 139, 144, 145, 162, 163, 172, 185, 190, 191, **192**-195, 201, 208, 282

Copper 22, 63, **94**, 95, 96, 104, 123, 154, 155, 186, 225, 246

Corundum 21, 22, 48, 186

Crystal(ize) 19, 22, 27, 29, 34, 35, 53, 55, 59, 61-64, 72, 74, 88, 96, 98, 100, 102-104, 112, 114, 115, 122-126, 136, 139, 151, 159, 188, 236, 237, 313

Crystalline 16, 28, 29, 65, 74, 75, 78, 8, 83, **116**, 117, 129, 165, 179, 186, 216, 217, 219, 221, 224, 226, 232

Crystalloblastic **115**

Crystallography 227, 266, 267

Crystal class 45-50, 84-86, 94

Crystal system 45-50, 84-87, 94, 96, **266**

Cummingtonite 123, 146, 275, 305

Cuprite 275

D

Dacite 60, 62, 64, 282

Datolite 22

Density 59, 94, 96, 165, 188, 214, 237, 241, **268**

Diamond 21, 22

Diamictite / diamicton 144, 191, 192, 206, 208-210, 282

Dickite 275

Digenite 275

Diorite 16, 37, 52, 79, 155, 282

Diopside 86, 88, 123, 154, 275

Dolomite 21, 22, 112, 115, 122, 123, 186, 212, 216, 225, 236

Dolostone 16, 139, 162, 211-213, 216, 217, 220, 221, 224, 225, 282

Donpeacorite 275

Dumortierite 275

E

Edenite 275

Eclogite 119, 120

Enstatite 87, 123, 275

Epidote 22, 46, 63, 85, 115, 121, 123, 125, 156, 159

Epsomite 222, 228, 275

Esseneite 275

Euclase 275

Euhedral 51, 78, 102, 103, 114, 115, 159

Eulite / ferrosilite 87, 275

F

Fayalite 123, 275

Feldspar 22, 23, 27-29, 33, 34, 35, 39, 42-45, 49, 50, 51, 53, 56, 57, 60, 62, 65-69, 73, 78-81, 99, 100, 103, 122, 123, 128, 158, 159, 186, 196-198, 230, 231, 306

Felsic 24, 30, 35, 52, 67, 69, 72-74, 80, 82, 108, 129, 130, 151-153, 232

Feldspathoid / foid 24, **33**, 34, 39, 49-51, 56, 57, 60, 62, 65, 66, 69, 88, 307

Ferrosilite / eulite 87, 275

Fluorite 21, 22, 50, 51, 186

Fossil(iferous) 139, 191, 212, 217, 219, 221, 227

Franklinite 275

G

Gabbro 35, 36, **52**, 53, 59, 71, 82, 86, 155, 282

Gahnite 275

Garnet 22, 115, 121, 124

Galaxite 275

Galena 275

Glaucochlorite 275

Glauconite 184, 231, 275

Glaucophane 121, 124, 125, 275, 305

Goethite 21, 22, 124, 187, 224

Gold 22, 94, **96**, 97, 104, 154, 178

Grainstone 217

Graywacke / greywacke 140, 201

Greenschist 120, 121, 126, 135-137, 139, 140

Gneiss 61, 108, 112, 113, 119, 124, 125, 127-130, **137**. 139, 140, 151, 153

Graphite 124, 276

Granite 24, 39, 52, 56, 70, 71, 82, 85, 88, 112, 129, 151, 209, 282

Granodiorite **33**, 70, 71, 78, 155, 282

Granophyre 71

Granulite 119, 120, 126

Gypsum 21, 22, 187, 222, 228

H

Halite / table salt 222, 276

Hematite 22, 112, 124, 146, 165, 187, 223, 224, 225, 232

Hercynite 276

Hornblende 21, 22, 45, 46, 47, 52, 53, 56, 57, 62, 72, 73, 78, 80, 84, 85, 98-100, 102, 104, 121, 124, 305

Hydrothermal 63, 78, 118-120, 139, 154, 255

Hypersthene 86, 87, 276

I

Ice 22, 23, 124, 187

Illite / illite group 206, 207, 276

Ilmenite 276

Iridescent 45, 47

Iridium 276

Iron 23, 28, 29, 52, 85, 87, 94, 97, 104, 122, 124, 154, 158, 184, 187, 225, 230

Ironite 113, 126, 133, **146**, 148

J

Jacobsite 276

Jadeite 87

Jasperlite 226

K

Kaolin / kaolin group 206, 207

Kaolinite 51, 276

Kirkite 276

Kosmochlor 276

Kyanite 115, 121, 122, 125, 127

L

Labradorite 23, 45, **47**, 70, 86, 306

Lamprophyre 17, 31, **39**, 40, 78, 79, 81, 86

Lapilli 75-77

Latite 60, 62, 65, 79

Lattice 29

Laumonite 121, 125, 276, 282

Lawsonite 115, 121, 125, 277

Lead (Pb) 23, 31, 154, 187

Leucite 23, 49, 50, **51**, 86, 307

Limonite 165, 187, 225

Limestone 64, 112, 139, 209, 212, 213, 216, 219-221, 224, 225

M

Mafic 16, 24, 30, 31, 35, 39, 52, 53, 56, 57, 67, 69, 70, 72-74, 80, 81, 84, 88, 98, 99, 129, 130, 151-153, 230, 232

Magnesite 187, **225**, 277

Magnetite 23, 46, 47, 52, 84, 115, 123, 125, 187, 224

Malachite 23, 94, 187, **225**, 277

Marble 16, 112, 113, 133, **139**, 142, 149, 282

Marl 282

Mica 23, 46, 62, 88, 123, 186, 187, 206

Microcline 23, 27, 42-45, 50, 51, 104

Molybdenite 23, 277

Monzodiorite **33**, 57, 69

Monzogabbro **33**, 69

Monzonite **33**, 282

Mudstone 113, 121, 135-137, 163, 185, 191, 198, 200-202, **204**-206, 208-210, 217, 227, 283

Muscovite 16, 21, 23, 29, 79, 80, 82, 115, 121, 187

Mylonite / mylonitic **129**

N

Natrolite 50, 51, 277

Nepheline 23, **49**, 50, 51, 102, 307

Nickel 97, 224, 277

Nosean / noselite 278

O

Oligoclase 23, 26, 28, 42, **46**, 70, 307

Olivine 21, 23, 26, 28, 29, 30, 35, 47, 48, 52, 56, 72, 80, **84**, 86, 88, 123, 125, 156, 184, 231, 305

Orthoclase 21, 23, 26-28, 29, , 42, 43, 44, 45, 51, 86, 104, 187, 305

Opal 278

Optics **269**

Orthopyroxene 84, 86, **87**, 121, 123, 159, 306

Oversaturated **24**, 25

Oxygen 28, 123, 158

P

Paarite 278

Packstone 217

Palladium 278

Pegmatite / pegmatitic 20, 61, 88, 98

Perovskite 23, 278

Phaneritic 20, 25, 31-35, 52, 55, 59, 60, 62, 66, 72, 78, 82, 98, 99

Phyllite 112, 119, 121, 126, 133, **135**, 136

Pigeonite 86, 88, 122, 126

Plagioclase 16, 22, 23, 26, 28, 29, 33, 34, **42**-45, 49-53, 56, 57, 61-63, 65-70, 72, 73, 79-81, 98, 99, 102, 103, 104, 121, 122, 156, 187, 196, 306, 307

Platinum 23

Prehnite 63, 115, 121, 126, 135, 140, 278

Proto quartzite / protoquartzite 128, 133, 140, 141

Pumpellyite 22, 23, 63, 121, 126, 135, 140

Pyrite 21, 94, **96**, 115, 188, 225, 228

Pyroclastic 31, **74**-77, 81, 83, 98

Pyroxene 23, 29, 30, 47-49, 52, 56, 72, 73, 78, 79, 80, 82, 84, 86, **87**, 98-100, 102, 115, 121, 156, 159, 184, **306**

Q

Quart 309

Quarter 309

Quartz 16, 21, 23-26, 28-30, 33-35, 39, 45, 46, 49, 52, 53, 56, 57, 62-69, 72-74, 79, 80, 82, 85, 95, 99, 100, 102, 103, 112, 118, 126, 128, 136, 142, 155, 157-159, 169, 170, 188, 196-198, 201, 224, 226, 230-232, 307

Quartz Alkali-feldspar Plagioclase, Feldspathoid (QAPF) 24, 31, **33**, 35, 52, 53-57, 59. 60, 64, 65, 57, 69, 70, 73, 98, 99, 10,197, 198

Quartz Feldspar Lithics (QFL) 196-199, 201

Quartzite 16, 112, 113, 133, 139-143, 158

Quartzolite / silexite 17, 24, **33**

R

Recrystallize 129, 151

Rhodochrosite 279

Rhodonite 154, 279

Riebeckite 126, 279

Rutile 115, 279

Rhyolite 24, 59, 60, 62, 64, **65**, 79, 82, 283

S

Sandite **140**, 141

Sandstone 34. 66, 112, 123, 136. 137, 140, 142, 162, 163, 166, 169, 185, 191, **196**-198, 200-204, 230, 232, 283

Sanidine 23, 27, 50, 51, 86, 307

Saturated **24**, 25, 30, 39, 73

Schist / schistose / schistosity 113, 119. 121, 124-129, 133, 135, **136**, 139, 140, 159

Shale 112, 119, 135, 167, 184, 185, 191, 204, **205**-207, 210, 219, 283

Siderite 23, 188, 224, 225

Silexite / quartzolite 17, 24, **33**

Silicon 28

Silt 19, 20, 117, 135, 140, 169, 173, 185, 200, 204, 205-210, 212, 217, 226, 283

Siltstone 135, 185, 191, 204, 205, 210,

Silver 23, 94, 154

Slate 64, 112, 119, 121, 131-133, **135**, 136, 201, 283

Specific gravity 45-51, 84-87, 94, 188, 206, 214, **268**

Sodalite 23, 49, **50**, 307

Sphene 23, 52, 115

Spinel 23, 154

Subhedral 48, 102, 114

Supersaturated **24**, 25

Syenite 16, **33**, 46, 49, 100, 155

T

Talc 21, 23, 115, 188

Tephra 75-77

Texture 15, 52, 59, 61-63, 79, 83, 109, 116, 129, 130, 165, 169, 194, 199

Thin section 269, 270

Tin 23

Titanite (sphene) 23, 52, 125

Tonalite **33**, 70, 71, 130,

Tonalite–trondhjemite–granodiorite (TTG) 70, 71, 94, 283

Topaz 21, 23, 188

Tremolite 23

Trondhjemite 70

Tuff 75-77, 103, 119, 171

U

Ultramafic 24, 31, 35, 39, 59, 74, 75, 78, 80. 88. 98, 225

Undersaturated 24, 25, 49, 50, 51

Unsaturated 24, 25, 30,

Uraninite 23, 158, 188, 280

V

Varve 204, 206, 210

Venite Migmatite 133, 151, 153

Vitreous 45-51, 84-87, 122, 124, 125

Volume   3, 15, 33, 53, 55, 65, 73. 109, 146, 170, 180, 201, 207, 237, 240, 248, 268, 309

W

Wackestone 217

Wentworth 19, 20, 59, 116, 117, 171, 173, 201, 216

Wollastonite 87, 115, 154, 280

X

Xenotime 280

Y

Yvonite 280

Z

Zeolite 63, 78, 119, 135, 139, 140, 280, 308

Zircon 23, 52, 188

The "index by subject" is not a complete consolidation of every thing in this book. I did my best to include all of the terms I feel are the most relevant. Nothing on p.252-265 are included in the above index, since the items on those pages are already in alphabetical order. Every thing on the " Basic Properties of Select Rocks and Minerals / Mineraloids" chart are in alphabetical order.

Made in the USA
Columbia, SC
09 April 2025

43834fee-5a3d-4d33-a9ee-1f2ce5d60c20R01